Wireless Mesh Networks - Security, Architectures and Protocols

Edited by Mutamed Khatib and Samer Alsadi

Published in London, United Kingdom

IntechOpen

Supporting open minds since 2005

Wireless Mesh Networks - Security, Architectures and Protocols
http://dx.doi.org/10.5772/intechopen.74910
Edited by Mutamed Khatib and Samer Alsadi

Contributors
Oladayo Olufemi Olakanmi, Adedamola Dada, J. Rejina Parvin, G. Merlin Sheeba, Faezeh Sadat S. Babamir, Murvet Kirci, Henrique M. Manuel Salgado, Rafael M.G. Kraemer, Luís M. Pessoa, Ichi Kanaya, Eri Itoh, Paulo Fernandes Fernandes da Silva Junior, Ewaldo Eder de Carvalho Santana, Mauro Sérgio Pinto Silva Filho, Alexandre Jean René Serres, Raimundo Carlos Silvério Freire, Paulo Henrique da Fonseca Silva, Maciel Alves de Oliveira, Fabrício Ferreira Batista, Elder Eldervitch Carneiro de Oliveira, Almir Souza, Severino Aires de Araújo Neto, Silva Neto, Carlos Augusto de Moraes Cruz, Vinh Truong Quang, Hoa Le Viet, Pavel B. Khorev, Hossein S. Ghadikolaei

First published in London, United Kingdom, 2020 by IntechOpen
IntechOpen is the global imprint of INTECHOPEN LIMITED, registered in England and Wales, registration number: 11086078, 7th floor, 10 Lower Thames Street, London, EC3R 6AF, United Kingdom
Printed in Croatia

British Library Cataloguing-in-Publication Data
A catalogue record for this book is available from the British Library

Additional hard and PDF copies can be obtained from orders@intechopen.com

Wireless Mesh Networks - Security, Architectures and Protocols
Edited by Mutamed Khatib and Samer Alsadi
p. cm.
Print ISBN 978-1-78985-203-5
Online ISBN 978-1-78985-204-2
eBook (PDF) ISBN 978-1-78985-485-5

For EU product safety concerns:
IN TECH d.o.o., Prolaz Marije Krucifikse Kozulić 3, 51000 Rijeka, Croatia, info@intechopen.com or visit our website at intechopen.com.

Meet the editors

Mutamed Khatib received his PhD degree in Wireless and Mobile Systems from USM in 2009. His research interests are in the field of mobile networks and coding. Since 2005, he has worked as an instructor in the Engineering Faculty at the Palestine Technical University (Kadoorie), Tulkarm, Palestine. He was the Head of Telecommunication Department for two years, the Dean of the Faculty of Engineering for four years, and he is now the VP for academics, and is teaching advanced courses in telecommunications and coding as an Associate Professor. Dr. Khatib has a number of publications in various international journals and conferences. He is also an author of books as well as the editor for both books and journals, where he also serves as a reviewer.

Samer Alsadi is an Associate Professor and he received his PhD degree in electrical power engineering from Moscow Power Engineering Institute (Technical University), Moscow, Russia, in 2000. He worked as a consultant in the Electrical Department of Jenin's municipality for one year, and since 2001, he has been working at the Palestine Technical University, Kadoorie, Tulkarm, in the Department of Electrical Engineering. His research interests focus on solar energy, power systems, protection systems, and load forecasting.

Contents

Preface

A wireless mesh network (WMN) is a wireless network with a large number of stationary wireless mesh routers that are connected using wireless communication techniques to form a mesh structure. Some of these routers act as a client wireless access point (such as laptops, PCs, and smart devices) with wireless connection where these routers attach themselves to the mesh network in order to transmit and receive data via the backbone mesh network. One or more routers are connected to the Internet and serve as gateways.

Nowadays, wireless mesh networks is a rapidly growing topic, but it is still largely new. After being used in military applications at the beginning, WMN moved to civil use and is now being used worldwide as both local area networks (LANs) and metropolitan area networks (MANs). Nevertheless, these arrangements are still 'cutting edge' and it is not yet clear what the most permanent applications of mesh will be – particularly as the application moves from early adopters towards wide-spread uptake.

This book discusses and investigates some issues related to WMNs, starting from a full overview of this system, describing, in depth, some related research and ending with interesting applications and supporting systems.

Mutamed Khatib, PhD and Samer Alsadi, PhD
Palestine Technical University - Kadoorie,
Tulkarm, Palestine

Chapter 1

An Overview of Wireless Mesh Networks

J. Rejina Parvin

Abstract

Wireless mesh networks (WMNs) are communication networks which comprise radio nodes in which nodes are arranged in a mesh topology. Mesh topology is an interconnection of all nodes connected with all other nodes in the network. The network includes devices like nodes, clients, routers, gateways, etc. As the nodes are fully connected, mesh networks are usually less mobile as rerouting is less difficult in predicting the reroute results in delay in data transmission. Mesh clients can be of any wireless devices like cell phones, laptops, etc. The gateways which act as forwarding nodes may not be connected with the Internet. As different devices come under a single network, it is also referred as mesh cloud. WMN is self-healable. It works better with various different networks which include cellular networks and IEEE 802.11, 802.15, and 802.16 as well. WMN is flexible to work with more than one protocol. This chapter gives architecture, layer functionalities, and applications.

Keywords: WMN architecture, layer functionalities, WMN standards, applications

1. Introduction

Wireless mesh network is a network which comprises various wireless nodes with access points. Each node in the network acts as a forwarding node to transfer the data. Since the network is decentralized, forwarding of data is possible only to the neighboring node. This results in the network structure simple and easy. WMN makes the people connected with the Internet who work at remote areas and operating business. This chapter throws light on WMN architecture, layer functionalities, various other networking standards, and applications.

2. Wireless mesh network

2.1 Architecture

Wireless mesh network is the architecture which provides less mobility with low cost within a radio range. WMN is an infrastructure which is a network of routers minus cabling between the nodes. It consists of radio nodes which need not to be cabled to a wired port like the conventional wireless access points. Shortest hops are predicted to transmit the data toward large distance [1]. Nodes between the source and destination act as a forwarding node which works cooperatively in making decisions in route prediction based on the topology and forwarding the data.

Wireless mesh network provides stability when compared to the rest of the network topologies rather than the node addition or deletion in the network. In infrastructure mesh network, the data forwarding and receiving are via gateway, whereas in the rest of the network, it is through pair of nodes [2].

The frequency of link breakage is higher in the case of wireless mesh networks when there is a high mobility which results in low performance in communication of information [3, 4].

Wireless mesh networks are categorized into three types based on the functionality of the nodes in the network:

- Infrastructure mesh architecture
- Mesh architecture based on clients
- Hybrid mesh architecture

2.1.1 Infrastructure mesh architecture

Mesh routers together act as a wireless back bone for infrastructure mesh architecture. Client node is passive in mesh infrastructure via Ethernet links; conventional clients with Ethernet interfaces can be connected to mesh routers.

If the traditional network and the mesh router are operating under the same radio range, then it is easy for the mesh network to communicate with the mesh router. Alternatively, if the radio ranges differ, the nodes will communicate to the base station so that with the help of Ethernet, it can be further communicated to the mesh routers. **Figure 1** shows the mesh architecture for infrastructure-based network.

2.1.2 Mesh architecture based on clients

Mesh architecture based on client is the one in which the client nodes are connected from peer to peer. Each node can act as a routing node to transfer the data. Here, the client performs the role of mesh routing by acting in the forwarding of the data packets.

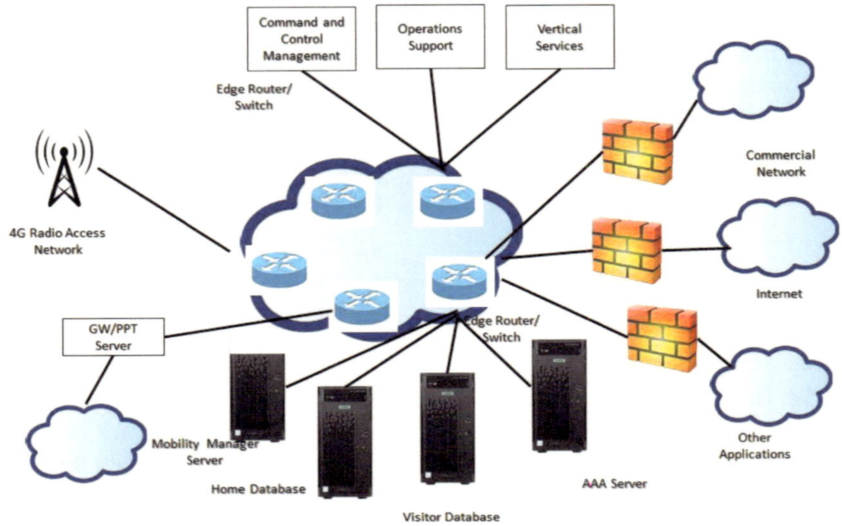

Figure 1.
Mesh architecture for infrastructure-based network.

Figure 2 shows the architecture of mesh based on clients. In this we can see that the network with no router is connected to it, and rather all clients are interconnected to perform data transfer.

2.1.3 Hybrid mesh architecture

In hybrid mesh architecture, usually the mesh nodes/router acts as a back bone of the entire network operation. With the help of network mesh router, it performs routing and forwarding of data packets toward its destination [5] which is shown in **Figure 3**.

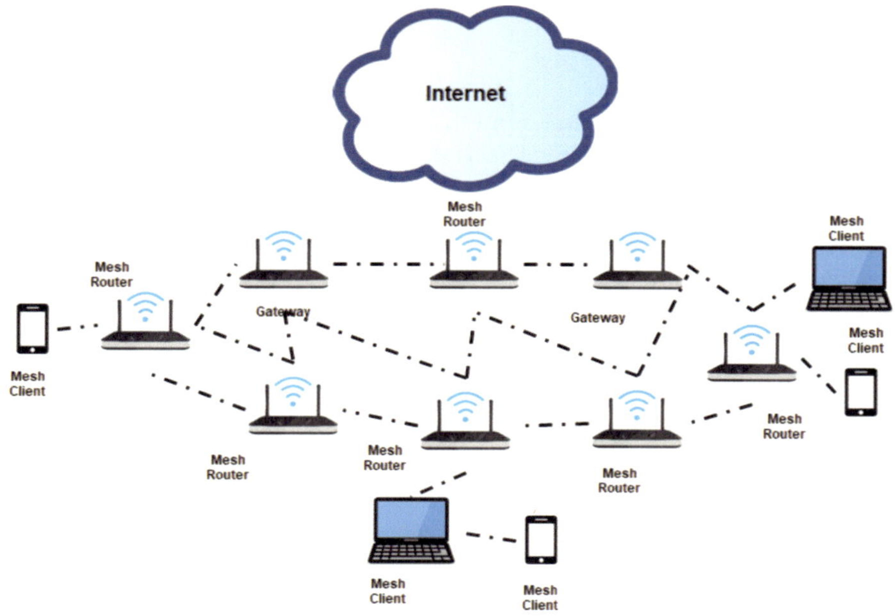

Figure 2.
Mesh architecture based on clients.

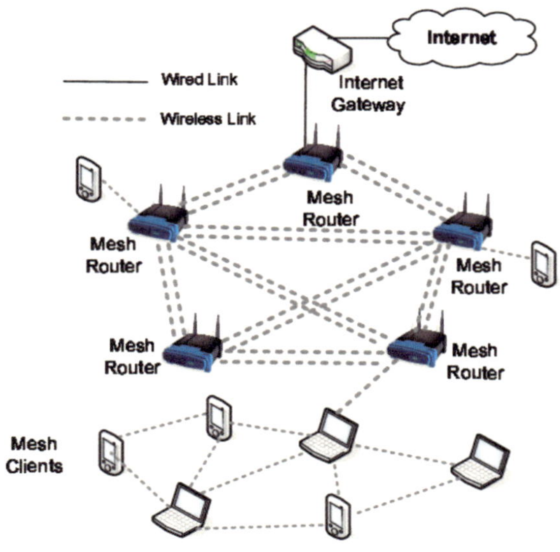

Figure 3.
Hybrid mesh architecture.

2.1.4 Characteristics of wireless mesh networks

- Dynamic self-configuration and self-organization
- Adaptation
- Fault tolerance and robustness
- Low-cost
- Integration and interoperability

2.2 Management

The infrastructure-based wireless mesh network is a decentralized network without a centralized management or with no centralized server which is more expensive. These methods are more reliable and efficient as each node has to transmit to the next node. Here, nodes act as router to transmit the data to its peers which are located far even it is a single hop. Wireless mesh network should be stable, i.e., there should not be high mobility. If node failure occurs due to any hardware problem or any other, the neighbor node will perform rerouting with the help of routing protocols.

2.2.1 Applications

Mesh network may comprise of mobile devices or stationary devices. Some of the applications of mesh networks which deserve communication are:

- Battlefield surveillance
- Tunnels
- Mobile video applications
- Emergency situations
- Tunnels
- Real time car racing, etc.

Voice over Internet Protocol (VoIP) is the main application of wireless mesh networks. In order to provide quality of service (QoS), wireless mesh network is used in telecommunication for voice communication. In current scenario, some of the applications where wireless mesh has been used include:

- Military forces in the USA are using wireless mesh networks to connect their devices for field operations.
- In residences, electric smart meters have been implemented to transfer the reading from one point to another (say as, from home to centralized office) to eliminate the man power.
- Wireless mesh network is used in one laptop per child program, in which it makes the students to share their files even when they are not connected with the Internet or with any physical connections.

- From 2010 onward, Wi-Fi-enabled mesh routers are available in the market which is installable even in homes or at small workplace. Google Home, Google on Hub, and Google Wi-Fi are various Wi-Fi wireless mesh networks.

- Iridium constellation with 66 satellites works under mesh network. This topology connects various wireless links with other satellite. Voice calls can be communicated via mesh networks between one satellite and another across constellations without transferring the signal to the ground station. Latency is highly reduced by avoiding the signal transfer to the nearby mesh instead of transferring to ground station.

2.2.2 Operation

The working principle of wireless mesh network is the same as that of the packets that travel around the wired internet data transfer between one node and another toward the destination [6]. This is implemented with the help of dynamic routing algorithm. It is possible by making each node communicate its routing information to other nodes within the network. With the received information, each node will decide whether to forward or to keep the data for itself. It is based on the functionality of the routing protocol. It is necessary for any routing algorithm to ensure that routing is done by predicting the shortest path between the source and destination. **Figure 4** shows the architecture of wireless mesh network.

Wireless mesh network can be connected with the existing network to provide effective communication. In traditional network comprises of various wired nodes, hotspots to communicate with the users, whereas in wireless mesh networks, the network is established with the help of various numbers of wireless nodes to communicate with each other.

The wireless nodes used in wireless mesh networks play same role as wireless routers. Various Wi-Fi communication IEEE standards like IEEE 802.11a, 802.11b, and 802.11 g are used for wireless mesh communication. Nodes in the network are

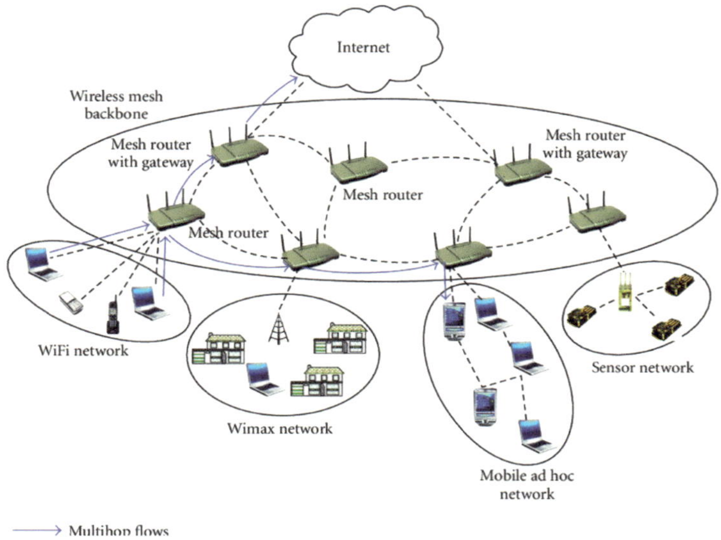

Figure 4.
Architecture of wireless mesh network.

capable of predicting the shortest path from the available path between source and destination. Addition and deletion of nodes and the routes will be updated then and there whenever there is a change in the network infrastructure. Dynamic routing is the capability of the node to predict the shortest available path between source and destination [6].

2.2.3 Advantages

- The cost of designing the network is lesser for fewer numbers of nodes even for large network coverage.
- Wireless mesh network shows better performance even for large number of nodes in the network.
- Wireless mesh networks relay on various Wi-Fi standards.
- It is useful for Non-line-of-sight (NLoS) network.
- It is self-configuring and self-healing.
- Easy to install and uninstall which makes network more adaptable with less or more number of nodes.

2.2.4 Backhaul nodes

In wireless LAN, there may be a chance that the information may be returned to the wired access point. Getting back the information to access is called backhaul. Small networks can be handled without any special configuration, whereas in larger network, backhaul nodes are required to retrieve the information from wired access node.

3. Protocol layer and functionalities

Factors that influence the performance of wireless mesh networks include:

- Architecture
- Topology
- Data pattern and traffic
- Density of the nodes
- Number of channels used by the nodes in the network
- Transmission power
- Mobility of the nodes

In order to develop the protocols, we need to clearly understand the relationship between the above factors and the capacity of WMNs and is given in **Figure 5**.

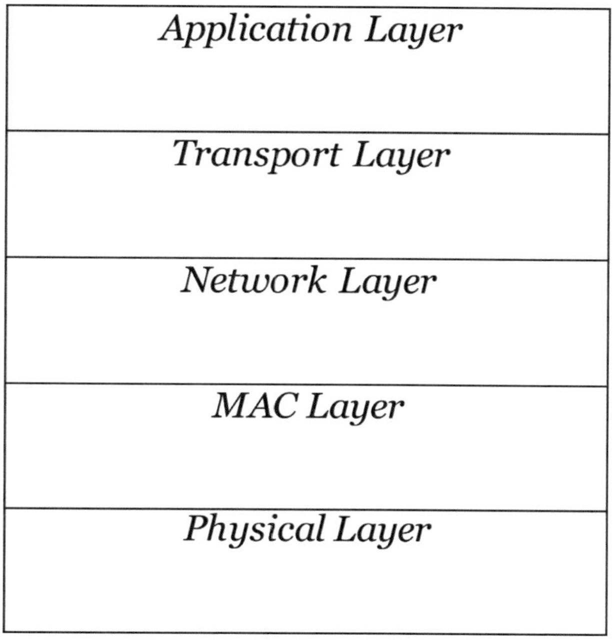

Figure 5.
Layer structure of WMN.

3.1 Physical layer

Wireless mesh networks are capable of multiple data rates simultaneously. It is achieved with the help of various modulation techniques and coding rates. Link adaption provides adaptive error resilience. High-speed data transmission is achieved with orthogonal frequency-division multiple access (OFDMA) and ultra-wideband (UWB) techniques.

Wireless communication system is provided with smart antenna for the purpose of increasing the capacity and to overcome the co-channel interference, fading, multi-antenna systems, etc. But it is tedious for designing the same for wireless mesh networks though it is available with the existing system. Unoccupied spectrum can be utilized by proper frequency planning with the help of WMN. Frequency agile/cognitive radio is used in order to utilize the unused spectrum.

As most of the radio components, RF band, various channel modulations, and access modes are programmable in such a way that it can be implemented in a software for working under cognitive radio range. It will be easier if further updating is to be carried out.

With the available physical test beds, the software platform is not much appropriate to provide desired solution. But still it can able to make advance changes in physical layer techniques which suit better for wireless communications.

3.2 MAC layer

MAC layer in WMN plays a unique role when compared to various other wireless networks:

- In WMN, communication is focused on more than one hop rather than single hop with the help of MAC.

- Multipoint to multipoint communication is possible with the help of distributed MAC.

- Self-healing and self-organization of node are the prime requirements of WMN for better performance.

- Communication between the node and the node which are located far (multihop distances) from the source node should be proper for providing better stability.

- Even less mobility influences the system performance of MAC.

- MAC protocol is designed in such a fashion that WMN can work simultaneously with both single and with multiple channels.

3.3 Routing layer

Though various routing protocols are available for ad hoc networks, it cannot be directly used for wireless mesh networks. It is still a big research to perform modifications of the available protocol to be adopted for WMN. There are the same salient features that WMN should provide with which are:

- Various performance metrics

- Scalability

- Robustness

- Energy-efficient routing algorithm suits for mesh network

Various routing algorithms in ad hoc network are available with any one of the above features, but not with all which makes it difficult to adopt with the mesh network. Some of the routing concepts for mesh network are as follows.

3.3.1 Multi-radio link quality source routing (MR-LQSR)

In MR-LQSR, a metric called weighted cumulative expected transmission time (WCETT) is used for measuring the system performance. In WCETT, link quality and minimum number of hop counts are considered as a system metric which results in better throughput and less delay.

3.3.2 Multipath routing

The objective of multipath routing is to provide load balancing and fault tolerance. Between the source and destination, multiple paths are predicted. When the shortest path link breaks, it can be easily switched over to other available paths. This results in better system performance by reducing the waiting time for computing the path at the time of fault. It also improves throughput and reduces end-to-end delay and fault tolerance. Still complexity exists with the multipath routing as it has to keep record of multipath always even if there is no breakage or fault.

3.3.3 Hierarchical routing

Hierarchical routing protocols show better performance when there is dense number of nodes in the network. This is due to fast setup procedure, reduced overheads, and shortest routing path. Complexity is higher on maintaining the hierarchy and directly relates to the performance of the system.

3.3.4 Geographical routing

It is a unique routing scheme in which it forwards the packets with the knowledge on the position of the node which is being communicated instead of the topology-based method. Geographical routing algorithm (single-path greedy routings) finds difficulty in delivering the data, if the path is available between source and the destination as data forwarding with the help of current location information.

Data delivery is guaranteed by using planar-graph-based geographic routing algorithms but results in more overhead information.

3.4 Transport layer

A large number of transport protocols are available for ad hoc networks, and WMNs depend on those transport layer protocols. Till date, there is no transport protocol that has been proposed specifically for WMNs. We know that ad hoc network is also not mature. It also has various unresolved issues. This suggests further research in this area.

3.5 Application layer

WMN supports enormous applications which include:

- Internet access
- Distributed information storage and sharing
- Information exchange across multiple wireless networks

Various research works have been carried out under these domains but focusing on modifying the existing application layer protocol by adapting various features for the mesh application layer protocol [7].

3.6 Issue in network performance

WMNs have their merits and drawbacks too. There are some issues to be concentrated for improving the network performance which include connectivity, radio range, interoperability, compatibility, etc.

3.6.1 Security issues

The weak area of WMN is security. Strong research is demanded due to the absence of centralized authority or key management for assuring security to provide fully trusted system.

3.6.2 Other issues

The common issues in WMNs are channel capacity expansion, scalability, and quality of service [8].

4. Conclusion

The nodes in WMN are self-configurable and self-healable. Such self- configuring nodes are better which improves the system performance, whereas self-healing makes the network to reconfigure if there are any addition and deletion of nodes in the network. Due to the huge number of nodes and data, there will be a high fault tolerance and degradation in performance. Integration of existing network leads to more complexity. By eradicating these drawbacks, the performance of WMN can be enhanced.

Author details

J. Rejina Parvin
Department of ECE, Sri Krishna College of Engineering and Technology, Coimbatore, India

*Address all correspondence to: parvinece@gmail.com

References

[1] Noel AB, et al. Abderrazak Abdaoui, Tarek Elfouly, Mohamed Hossam Ahmed, Ahmed Badawy and Mohamed S. Shehata. Structural health monitoring using wireless sensor networks: A comprehensive survey. IEEE Communications Surveys & Tutorials. 2017;**19**(3):1403-1423

[2] Chen SM, Lin P, Huang DW, Yang SR. A study on distributed/centralized scheduling for wireless mesh network. In: Proceedings of the 2006 International Conference on Wireless Communications and Mobile Computing. British Columbia, Canada: Vancouver; 2006. pp. 599-604

[3] Pathak PH, Dutta R. A survey of network design problems and joint design approaches in wireless mesh networks. IEEE Communications Surveys & Tutorials. 2011;**13**(3):396-428

[4] Alanazi S, Saleem K, Al-Muhtadi J, Derhab A. Analysis of denial of service impact on data routing in mobile eHealth wireless mesh network. Mobile Information Systems. 2016;**2016**:19. DOI: 10.1155/2016/4853924. Article ID: 4853924

[5] Porto DCF, Cavalcanti G, Elias G. A layered routing architecture for infrastructure wireless mesh networks. In: Fifth International Conference on Networking and Services, 2009. ICNS '09: 2009. pp. 366-369. DOI: 10.1109/ICNS.2009.91 [Retrieved: 14 November 2016]

[6] Ashwood-Smith P. Shortest Path Bridging IEEE 802.1aq Overview (PDF). Huawei. Archived from the original (PDF) on 15 May 2013. 2011. [Retrieved: 11 May 2012]

[7] Terence D. Todd A, Sayegh A, Mohammed NS, Dongmei Z. The need for access point power saving in solar powered. WLAN Mesh Networks Archived 2009-05-26 at the Wayback Machine. In: IEEE Network. May/June 2008

[8] Jun J, Sichitiu ML. The nominal capacity of wireless mesh networks. IEEE Wireless Communications. 2003;**10**(5):8-14

Chapter 2

Wireless Sensor Networks (WSNs): Security and Privacy Issues and Solutions

Oladayo Olufemi Olakanmi and Adedamola Dada

Abstract

Wireless sensor networks (WSNs) have become one of the current research areas, and it proves to be a very supportive technology for various applications such as environmental-, military-, health-, home-, and office-based applications. WSN can either be mobile wireless sensor network (MWSN) or static wireless sensor network (SWSN). MWSN is a specialized wireless network consisting of considerable number of mobile sensors, however the instability of its topology introduces several performance issues during data routing. SWSNs consisting of static nodes with static topology also share some of the security challenges of MWSNs due to some constraints associated with the sensor nodes. Security, privacy, computation and energy constraints, and reliability issues are the major challenges facing WSNs, especially during routing. To solve these challenges, WSN routing protocols must ensure confidentiality, integrity, privacy preservation, and reliability in the network. Thus, efficient and energy-aware countermeasures have to be designed to prevent intrusion in the network. In this chapter, we describe different forms of WSNs, challenges, solutions, and a point-to-point multi-hop-based secure solution for effective routing in WSNs.

Keywords: wireless sensor network, encryption, routing protocol, security, privacy

1. Introduction

Wireless sensor network (WSN), as shown in **Figure 1**, is a wireless interconnected network which consists of independently setup devices that monitor the conditions of its environment using sensors. WSNs are employed in a wide range of applications such as security surveillance, environmental monitoring, target tracking, military defense, intrusion detection, etc. Security in wireless sensor network is at a growing stage mainly not because of nonavailability of efficient security schemes, but most of the existing schemes are not suitable due to the peculiarity of WSNs. That is, WSNs' nodes have low computational capacity and energy constraint. In WSNs, sensor nodes have the ability to communicate with one another, but their primary task is to sense, gather, and compute data. These data are forwarded, via multiple hops, to a sink which may use it or relay it to other networks. To achieve an effective communication, WSNs need efficient routing protocols [2–6]. They facilitate communication in WSNs by discovering the appropriate routes for transmitting data and maintain the routes for subsequent

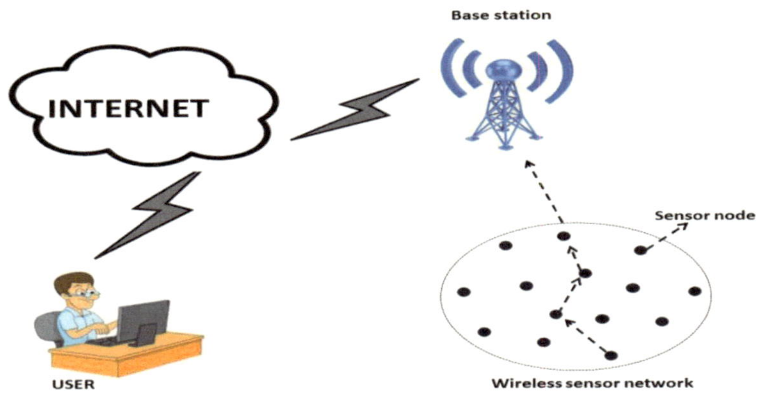

Figure 1.
A typical wireless sensor networks (WSN) [1].

transmissions. As a result of heterogeneity of WSNs' nodes, different protocols had been developed for different WSNs depending on the nature of the nodes and application. For instance, there are dedicated protocols for MWSNs and dedicated protocols for SWSNs.

There are two modes of transmission in WSN; single hop involves the source node sending its data packets to the destination within a hop. Meanwhile, WSNs' sensor nodes may rely on one another in order to relay packets to remote destinations. This mode of transmission is called multi-hop. Multi-hop is a routing phenomenon that involves the transfer of data between source and destination nodes with the cooperation of intermediary nodes. It enhances the performance of WSNs by allowing energy-depleted node to transfer data through its neighboring nodes along the routing path to the destination node. There are several security and privacy issues associated with multi-hop routing. Some of these issues like snooping, sinkhole, tampering Sybil, clone, wormhole, spoofing, etc. affect the integrity, availability, and data confidentiality of the WSNs.

Several security solutions had been proposed for WSNs; however, resource constraint of sensors makes some of these security solutions unfit for WSNs. This, therefore, makes their adoption in WSNs impossible. This is as a result of instability of the topology of most WSNs. Some of the WSNs, unlike some other networks, consist of mobile nodes that intermittently change the topology of the networks, therefore making it impossible for such mobile network to use existing protocol developed for static nodes. Also, large volume of data is transferred on the WSNs; this increases the traffic on the wireless communication infrastructure of WSN. All these show that security and privacy solutions of WSN must not only be lightweight in terms of the computational, communication, and energy overheads but also support aggregation and multi-hop in order to reduce the traffics and extend the life span of the networks. Meanwhile, most of the existing security solutions do not have these performance requirements [1, 7–10].

2. Classification of WSNs protocols

Routing protocols can be classified into:

1. Data-centric routing protocol

2. Hierarchical routing protocol

3. Multipath-based routing protocol

4. Location-based routing protocol

5. QoS-based routing protocol

6. Mobility-based routing protocol

2.1 Data-centric routing protocol

Data-centric routing protocol combines data arriving from various sensor nodes at a specific route. This eliminates redundancies and minimizes the total amount of data transmission before forwarding it to the base station. Directed diffusion, rumor routing, and sensor protocol for information via negotiation (SPIN) protocol are examples of data-centric routing protocol [11, 12].

SPIN is a negotiation-based data-centric protocol for WSNs. Each node uses metadata to name its data, and negotiation is performed by a sensor node using its metadata. Hence, each node is able to negotiate whether to deliver data or not, in order to eliminate redundant data transmission throughout the network. After the negotiation, the sender transmits its data as shown in **Figure 2**; node A starts by broadcasting its hop request to its neighboring node B. Once the request is accepted, node A sends its data to B who then repeats this procedure. This is to find its neighboring node and hops the data to the neighboring node until the data reaches the destination. SPIN protocol saves energy due to the fact that each node only performs single hop. SPIN's hop request and acceptance packets prevent flooding attack on WSNs. Although SPIN protocol is good for lossless networks, it can also be used for lossy or mobile networks.

2.2 Hierarchical routing protocol

Hierarchical routing protocol classifies network nodes into hierarchical clusters. For each of the clusters, the protocol selects a node with high residual energy as the cluster head. The sensed data of each node in the cluster are transferred through the cluster heads of the clusters in the network [11]. The cluster node aggregates the sensed data of all the nodes in the cluster before sending it to the sink. Hierarchical routing protocol reduces the energy consumption through multi-hop transmission mode [13]. Also, data aggregation performed by the cluster head reduces traffic on

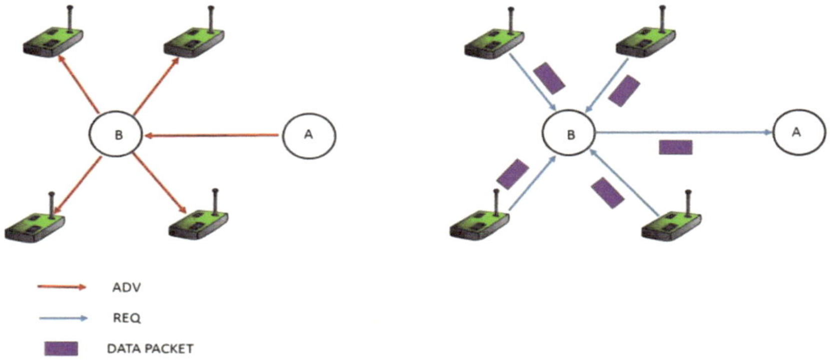

Figure 2.
SPIN protocol.

the network. Low-energy adaptive clustering hierarchy (LEACH), threshold-sensitive energy-efficient sensor network protocol (TEEN) and adaptive threshold-sensitive energy-efficient sensor network protocol (APTEEN), and secure hierarchical energy-efficient routing (SHEER) are examples of hierarchical routing protocol. TEEN gives a very good performance since it reduces the number of transmissions [14]. Patil et al. presented SHEER in [15]. It uses adaptive probabilistic transmission mechanism for determining the optimal route in WSN. SHEER also adopts hierarchical key establishment scheme (HIKES) for key distribution, authentication, and confidentiality. SHEER involves four phases as described below:

2.2.1 Initiation phase

1. The base station (BS), computes key $K_R = HMAC(I_R \| O_R)$, generates a broadcast authentication token N_R, and encrypts it as $N_R^l = Enc_{K_R}\ (N_R)$. The base station pre-loads each sensor node with N_R^l and keeps I_R and O_R.

2. BS broadcasts the initiation call as $N_b \parallel I_R \parallel O_R \parallel Enc_{K_R}\ \left(init \parallel N_b \parallel N_R \| N_R''\right)$, where *init* is the initiation call, O_R is the index, O_R is the offset of K_R, and N_b is time stamp generated by BS.

3. On receiving the initiation message, the sensor node extracts and decrypts $Enc_{K_R}\ \left(init \parallel N_b \parallel N_R \| N_R''\right)$, regenerates N'_R, and compares it with the N_R in the received initiation message. If they are similar, then the base station is successfully authenticated. It then replaces N_R in the newly with N'_R, sets its timer and starts the next phase.

2.2.2 Neighbor discovery phase

During the neighbor discovery phase, the sensor nodes establish their hopping link with their neighboring node. Each node switches from listening mode to transmission mode. In listening mode, node sends a HELLO message containing its identity, a nonce, and an encrypted header with the sensor key until it gets a reply from its neighboring nodes.

2.2.3 Clustering phase

In this phase, cluster consisting of certain number of nodes with a cluster head is selected based on some parameters.

2.2.4 Data message exchange phase

Each sensor sends its data to the base station through the cluster heads. This centralize data transmission reduces collision within clusters.

2.3 Multipath routing protocol

For an effective data delivery, multipath routing protocol generates a multipath (primary and secondary paths) from the source node to the destination node. It uses secondary path in case the primary path fails. With this, fault tolerance is achieved. However, this increases the cost of routing through the cost of maintaining multiple paths between source and destination [10, 16]. There are different types of multipath-based routing protocols.

2.3.1 Disjoint path routing protocol

In a disjoint path routing protocol, every source node finds the shortest disjointed multipath to the sink node. It evenly shares its data load among these disjointed paths. All the paths in this multipath share no sensor node. The protocol is reliable with extra overhead but at a low energy.

2.3.2 Braided path routing protocol

To construct braided multipath, the protocol first selects the primary path; then for every sensor, the best path is chosen from source to sink node, but this path does not include the primary node. The best alternative paths that are not necessarily disjoint from the primary path are called idealized braided multipath. These alternative paths are located either on the primary path or very close to it which means that the energy consumption on both the primary path and an alternative path is almost equal [17].

2.3.3 N to 1 multipath discovery routing protocol

N to 1 multipath discovery protocol is a protocol based on flooding. Example of N to 1 multipath-based routing protocol is multipath-based segment-by-segment routing (MSSR) protocol proposed by Lu et al. in [18]. MSSR protocol divides a single path into multiple segments, where multiple node-disjoint paths are discovered and independently maintained. N to 1 multipath discovery routing protocol reduces congestion, and effectively manages.

2.4 Location-based routing protocol

Location-based routing protocol routes data based on the distance of the source and destination nodes. It calculates the distance between source and destination nodes in order to determine estimated routing energy. Shruti [19] proposed a location-based routing protocol. The protocol uses the signal strength of the incoming signal to determine their distance. In their protocol, all the non-active nodes are put in sleeping mode in order to save energy. In location-based, the knowledge of the position of sensor nodes is exploited to route the query from the base station to the event. Location information enables the network to select the best route.

Another example of the location-based protocol is the geographic adaptive fidelity (GAF) protocol for mobile adhoc networks (MANETs). GAF conserves energy, and reduces routing overhead, which makes suitable for WSNs. Other examples of location-based protocols are location-aided routing (LAR), energy-efficient location-aided routing (EELAR), greedy location-aided routing protocol (GLAR), etc.

2.5 Quality of service (QoS)-based routing protocol

QoS-based routing protocol balances effective data delivery of the data to the sink node with some predetermined QoS metrics [17, 20]. Some of the existing QoS-based routing protocols are described below:

2.5.1 Sequential assignment outing (SAR) protocol

SAR protocol uses energy, QoS on each path, and the priority level of each packet as the QoS metrics to achieve effective data delivery. SAR protocol discovers and uses multiple paths from the sink node to sensor nodes for effective data delivery. SAR protocol considers energy efficiency and fault tolerance and also focuses on minimizing the average weighted QoS metric during data transfer [21].

2.5.2 SPEED protocol

SPEED is also an example of QoS-based routing protocol. In SPEED, every sensor node keeps its neighboring node information in order to increase the performance of the protocol. For example, SPEED protocol has congestion avoidance mechanism that is used to avoid congestion. The mechanism relies on the node information. Routing module in SPEED is called stateless geographic nondeterministic forwarding (SGNF) and works together with four modules at the network layer. In this protocol, the total energy used for transmission is incomparable to the performance of the routing algorithm.

2.5.3 QoS-aware and heterogeneously clustered routing (QHCR)

It is an energy-efficient routing protocol used by heterogeneous WSNs for delay-sensitive, bandwidth-hungry, time-critical, and QoS-aware applications. The QHCR protocol provides dedicated paths for real-time applications as well as delay-sensitive applications at a lower energy. The QHCR protocol consists of information gathering, cluster head selection, and intra-cluster communication phases.

2.6 Mobility-based routing protocol

Mobility-based routing protocol is a lightweight protocol that ensures data delivery from source to destination nodes. Tree-based efficient data dissemination protocol (TEDD), scalable energy-efficient asynchronous dissemination (SEAD), two-tier data dissemination (TTDD), and data MULES are some of the examples of mobility-based routing protocol. These routing protocols deal with the dynamism of the topology of the network. The closest node to the sink node tends to transmit more than others, which reduces its lifetime faster than other nodes [22]. Another example of the mobility-based routing protocol was the protocol proposed by Kim et al. [23]. The authors proposed a temperature-aware mobility algorithm for wireless sensor networks. Their algorithm employs store-and-carry mechanism to overcome the challenges posed by human postural mobility. In their store-and-carry-based routing protocol, routing packets are stored in a temporary memory called buffer. The buffer reroutes lost data to any intermediary node that temporarily lost connection with the source node. Their protocol also uses temperature to determine the intermediary node.

Another example of mobility protocol is the routing protocol proposed by Kumar et al. in [24]. They use ant colony optimization (ACO) and endocrine cooperative particle swarm optimization (ECPSO) algorithms to enhance the performance of the WSNs.

3. Security and privacy issues in WSN

Most of the existing WSN routing protocols and existing security solutions are unsuitable for WSNs. This is due to resources constraint associated with WSNs [25]. These constraints majorly determine the kind of security approaches that can be adopted for WSNs. Various security issues and their solutions are described in this section.

3.1 Security and privacy issues

The increase in demand for a real-time information has made WSN become more expedient. WSNs most of the time employs multi-hop transmission mode to overcome their constraints. The major problem of multi-hop transmission is attacks on the source data and nodes' identities during hopping. For a resource-constraint WSN with source node sending data to the destination through several intermediary nodes, there is a possibility of intrusion, identity tracing by an adversary, gleaning, and modification of source data by the intermediary nodes. WSNs, most times, operate in hostile environments and can be subjected to side channel attacks, such as differential power analysis. In these attacks, the adversary monitors the system, repeats the same operation, and takes careful measurements of power consumed in a cycle-by-cycle basis in order to either recover the secret key or perturb used in the perturbation. To prevent this, a scalar blinding is usually engaged in cryptographic-based security solutions. The scalar multiplication is blinded using integer m, where m is the order of the point $P \in E_q$, such that $mP = 0$. For example, instead of computing $Q = kP \bmod q$, $Q = (k + m)P \bmod q$ is computed.

Another issue in WSNs is how to preserve the identities of the source and destination nodes from the privy of intermediary nodes and adversaries during multi-hop. That is, there must be a form of lightweight authentication feature(s) inherent in the data packet between a source and destination nodes. Some other attacks on WSNs are discussed below.

3.1.1 Manipulating routing information

This attack targets the routing information between two sensor nodes. It can be launched through spoofing or replaying the routing information. This can be done by adversaries who have the capability of creating routing loops, attracting or repelling network traffic, and extending or shortening source routes. This attack is a passive attack which is not only easy to launch but elusive to detection. However, a unique identity can be created for the selected path (using key-based hash function of the pseudonyms or identity of all the selected intermediate nodes and embellishes in the message, any attempt to record data packet from a location and re-tunnel it at another location will be detected by the base station when comparing the embellished path identity with hash of all the appended pseudonyms or identities of all the nodes involved in the multi-hop).

3.1.2 Sybil attack

In this attack, adversary compromises the WSN by creating fake identities to disrupt the network protocols. Sybil attack can lead to denial of services. It may also affect mapping during routing, since a Sybil node creates illegal identities in a bid to break down the one-to-one mapping between each node. Sybil is common in P2P networks and also extends to wireless sensor networks [8]. Moreover, detection and defense against Sybil attack is more challenging; this is due to the limited energy and computational capabilities of WSNs. Different efforts had been developed to thwart Sybil attack in WSN. An example is the use of a pair-wise key-based detection scheme which sets a threshold for the number of the identity that a node can use [21]. However, this requires pre-assignment of keys to sensor node.

Another way to thwart Sybil attack is to validate identity of every node involved in routing. This can be reactively or proactively done. Reactively means prior to routing, a node must provide enough identification parameters to differentiate it from all other sensor nodes. The most common method is a resource test. Another way is to increase the cost against the benefit in identity generation [8]. That is, increasing cost of creating an identity and reducing the possible of having multiple identities will thwart Sybil attack, since the goal of a Sybil attacker is to acquire more identities. Also, traceable pseudonym and network-node identity generated by base station can be used to prevent a Sybil attack [9, 26].

3.1.3 Sinkhole attack

This attack prevents the sink node (base station) from obtaining the complete and correct data from the sensors, thus posing a threat to higher layer applications. In this attack, an adversary makes itself receptively attractive to its neighboring nodes in order to direct more traffics to itself [27, 28]. This results in adversary attracting all the traffics that is meant for the sink node. The adversary can then launch a more severe attack on the network, like selective forwarding, modifying, or dropping the packets. WSN is more vulnerable to this attack because its nodes most of the time send data to the base station [29].

Meanwhile, a point-to-point authentication between source node, identifiable intermediate nodes, and end-to-end symmetric encryption between source and destination nodes can be used prevent sinkhole, Sybil, and sinkhole attacks. The attack is foiled once the adversary could not decrypt end-to-end symmetric encrypted data even if it successfully impersonates the node and receives its data packet [9].

3.1.4 Clone attack

In a clone attack, the attacker first attacks and captures the legitimate sensor nodes from the WSNs, collects all their information from their memories, copies them on multiple sensor nodes to create clone nodes, and finally deploys them to the network. Once a node is clone, adversary can then launch any other attacks. There are two different ways of detecting this attack: centralized and distributed approaches. Centralized uses sink node to detect and foil the activities of clone nodes, while distributed approach uses selected nodes to detect clone nodes and foil their activities in the network. Distributed approach is suitable for static WSNs because distributed techniques use nodes' location information to detect clones and sensor nodes with the same identity, but different addresses are taken as clone nodes. Meanwhile, in mobile WSNs, it is a different thing entirely, sensor nodes keep changing their position, and these nodes keep joining and leaving the network. Hence, node location information is not considered as the best technique for detecting clone nodes. Clone node can launch the following attacks:

3.1.4.1 Selective forwarding attack

Multi-hop-based WSN routing protocols assumed that all the neighboring nodes must re-hop their received data packets. Malicious nodes selectively forward some packets while dropping the others. Selective forwarding attacks are most effective when the adversary is actively involved in the data flow.

3.1.4.2 HELLO flood attack

This attack utilizes the connection between nodes. Most routing protocols require sensor nodes to broadcast HELLO packets to announce themselves to their neighboring nodes. An adversary may exploit this to deceive sensor nodes receiving the HELLO packet that they are within the radio range of the source node. In [30], the authors proposed a new method for detecting the HELLO flood attack based on distance. Here, nodes not only compare the RSS of the received HELLO packet but also compare the node's distance to the selected cluster head (CH) with the threshold distance. Only those nodes whose RSS as well as distance falls within the threshold limits are allowed to join the network. For example, in the setup phase of LEACH protocol [31], CH sends its own location coordinates. The nodes receiving HELLO packets from CH calculate the distance *Dist* as shown below:

$$Dist = sqrt[sq(x2 - x1) + sq(y2 - y1)]$$

Here, (x1; y1) are the coordinates of the sensor node receiving the packet, and (x2; y2) are the coordinates of CH. Each sensor node calculate the radio signal strength value (*RSS*) and distance between (*Dist*). These are used to determine the cluster they belong in, that is, if $(RSS < ThRSS \text{ and } Dist < ThDist)$ then Node = 'Friend of the cluster' otherwise not a friend of the cluster.

3.1.4.3 Denial of service attack

This type of attack exploits the weaknesses in the sensor network, by attempting to disrupt the sensor network. Denial of service (DoS) attack denies services to valid users [32]. In a safety-critical network, this kind of attack can be disastrous to the functionality of the network. One of the methods engaged by adversary to launch DoS is by flooding the network with messages in order to increase traffics on the network. The DOS attack can be detected through proper filtration of incoming messages based on the contents and identifying nodes with high number of faulty messages. Faulty messages are detected by checking for the contradiction between messages sent by neighboring nodes [33].

3.2 Security and privacy solutions

Recently, application of WSN has gained massive attention leading to new security challenges and design issues [34]. In this section, we discussed relevant research efforts on the development of security schemes for WSN using different approaches such as effective key management, public key infrastructure (PKI), multiclass nodes, as well as grouping of nodes to improve the security of routing protocols in WSNs.

3.2.1 Use of effective key management

Du et al. presented a scheme with an example of an effective key management. Their scheme takes advantage of the high-end sensors in the heterogeneous networks. The performance evaluation and security analysis of their scheme show that the key management scheme provides better security with less complexity than the existing key management schemes [35]. The protocol pre-assigns a few keys in the L-sensor and a few keys to every H-sensor. This is because H-sensor is tamper-proof and has a larger memory than L-sensor. Their scheme uses asymmetric pre-distribution (AP) key management scheme since the number of pre-distributed keys in an H-sensor and in an L-sensor is different [12].

3.2.2 Use of effective public key infrastructure

Yu in [36] solved the security problem in WSN using the public key cryptography as a tool to ensure the authenticity of the sink node or base station. The approach consists of two phases; the first phase is node to sink handshake phase, where sink and sensor nodes set up session keys for secure data exchange. In the second phase, the session keys are used to encrypt data. Their scheme is very easy to implement, and requires a low computational power. The only limitation of their scheme is that all the participating nodes in the network have to agree on a common key prior to the exchange of data. However, any scheme based on a single key is vulnerable to the key compromise. That is, a compromised sensor node will not only compromise the shared key but also the whole network.

Also, Chen et al. [37] presented a PKI-based approach to ensure secure keys exchange in the WSNs. Their scheme provides key management mechanism for wireless sensor network applications that can handle sink mobility and deliver data to neighboring nodes and sinks without failure. They also presented a method for detecting and thwarting DoS attack and data authentication encryption.

3.2.3 Effective use of multiclass nodes

Du et al. [38] presents a new secure routing protocol for heterogeneous sensor networks (HSNs), which is a two-tier secure routing (TTSR) protocol. The TTSR protocol consists of both intra-cluster routing and inter-cluster routing schemes. The intra-cluster routing forms a minimum spanning tree (shortest path tree) among L-sensors in a cluster for data forwarding. In case of inter-cluster routing, data packets are sent by H-sensors in the relay cells along the direction from the source node to the sink node. The tree-based routing and relay via relay cells of TTSR make it resistant to spoofing, selective forwarding, and sinkhole and wormhole attacks.

Du [39] also proposed a novel QoS routing protocol that includes bandwidth calculation and slot reservation for mobile ad hoc networks (MANETS). Their QoS routing protocol takes advantage of the numerous transmission ability of multi-class nodes. Their protocol used three encryption keys:

1. A public key known by the sink and all other nodes

2. Node private key shared by two neighbor nodes and refreshed in the route discovery phase

3. A share primary key between node and sink node

The QoS routing protocol divides transmission data into different data slices. Each slice is route through a unique route of the discovered multipath.

3.2.4 Effective grouping of nodes to improve security of wireless sensor networks

In group-based WSN security scheme, the dominating node processes the sensed information locally and prepares the authenticated report for the destination node [40]. In this category, sensor nodes are grouped into smaller clusters wherein each cell assigns a special sensor node to carry out all the burden of relaying multi-hop packets. Hence division of labor is possible in the network, which makes the scheme to consume low power. Zhang et al. in [41] presented a group-based security

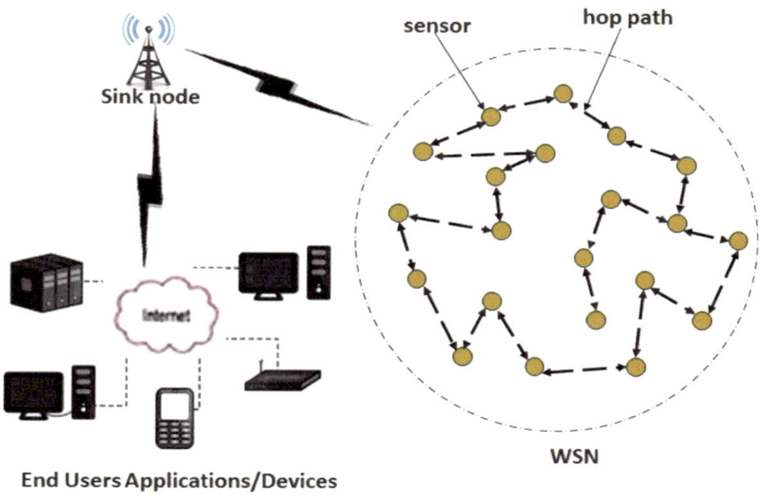

Figure 3.
Wireless sensor network system model.

scheme for distributed wireless sensor networks; their scheme involves three entities: one or more sink nodes, Y number of group dominator nodes, and N number of ordinary sensor nodes.

3.2.5 Point-to-point security solution

Point-to-point security solution involves secure routing between every two nodes along the multi-hop path. To show the design and efficacy of point-to-point solution, we fully describe a typical point to point security solution for multi-hop based WSNs proposed in [9]. Olakanmi and Dada [9] proposed an effective point-to-point security scheme that engages point-to-point (PoP) mutual authentication scheme, perturbation, and pseudonym to overcome security and privacy issues in WSNs. To reduce computational cost and energy consumption, they used elliptic curve cryptography, hash function, and exclusive OR operations to evolve an efficient security solution for a decentralized WSNs. The network model, as shown in **Figure 3**, consists of base station (BS), immediate node (IN), source node (SN) or (sn), and destination node (DS) or (ds). The SNs and DSs are capable of multi-hop transmission; therefore any SN can become DS and vice versa.

The PoP security scheme consists of the following phases: registration and key management, secure data exchange, perturbs generation, signature and obfuscation, authentication, and verification and decryption phases.

3.2.5.1 Registration and key management phase

The serial number ψ of each node is sent to BS. BS then generates unique pseudonym and network-node identity as follows:

i. BS randomly generates $s, \rho \in Z_q^* *$, as its master secret key pair, and computes and distributes its public parameter $\varphi = (\rho + \mu)P \ mod \ q$, where P is the generator of elliptic curve E_q and q is the order of E.

ii. Each node i randomly selects a unique $r_i \in Z_q^*$, computes its two-way distribution parameter β_i as $\beta_i = (r_i + \mu)P \ mod \ q$, and broadcasts its β_i to other nodes in the network.

iii. BS then computes N_i as $N_i = H(\rho \oplus \psi_i)$ and pseudonym F_i for each registered node as $F_i = H(N_i \| s \| \psi_i)$. It extracts the distribution parameter β_i of the node i in order to compute its node-base station shared key $\gamma_{bs \to i}$ as $\gamma_{bs \to i} = \rho \ \beta_i$ and sends the symmetrically encrypted node's F_i and N_i as $E_{\gamma_{bs \to i}}$ (F_i) to node i.

iv. On the receipt of its encrypted pseudonym, each node then generates its corresponding node-base station shared key as $\gamma_{i \to bs} = r_i \varphi$ and uses it to decrypt the received encrypted pseudonym.

3.2.5.2 Secure data exchange phase

To send data M, the primary SN signs M and generates perturb to secure M. It then encrypts the obfuscated message packet as σ, using its node-destination shared key $\phi sn \to ds$. The message packet σ contains the signature δ, perturbed data P_p, pseudonyms of the primary source node F_{sn}, and destination node F_{ds}.

3.2.5.3 Perturb generation phase

The perturbation enforces first level of security on the data. It is used to remove semantic pattern caused by wide variation in the transmitted data. The perturbation uses a novel additive noise generation method to perturb the data M. Primary source and destination nodes independently generate a set of perturb λ for session τ as follows:

i. The SN and its destination node generate their perturbation parameters α_{sn}, α_{ds} by randomly selecting a unique $m_1 \in Z_q^*$ and $m_2 \in Z_q^*$, and compute $\alpha_{sn} = (m_1 + \mu)P \ mod \ q$ and $\alpha_{ds} = (m_2 + \mu)Pmod \ q$, respectively.

ii. Using the destination perturbation parameter α_{ds} for session, SN computes perturbation seed ϑ as $\vartheta = m_1 \alpha_{ds}$.

iii. For session, SN generates the perturbation chain as $\lambda = \{\lambda_1, \lambda_2, \lambda_3 \ldots \lambda_k\}$, where $\lambda_1 = H_\vartheta(\vartheta \| F_{sn})$, $\lambda_n = H_\vartheta(\lambda_{(n-1)})$ for $n = 2 \ldots k$. Clear all the perturbation parameters of perturb index $n - 1$ in its memory for session τ and destination node of pseudonym F_{ds}. It replaces its former encrypted perturbation parameters with the new one, that is, replaces $[(\lambda_{n-1} \| m_1 \| n - 1 \| F_{ds}) \oplus \vartheta]$ with $[(\lambda_n \| m_1 \| n \| F_{ds}) \oplus \vartheta]$.

iv. Primary SN computes new perturb for every new data transmission of the same session by repeating step c using the previously used perturb λ_{n-1}. However, for a new session and destination node, SN generates a new ϑ by following steps (i)-(iii).

3.2.5.4 Signature and perturbation phase

Primary source node signs and perturbs the data packet through the following process:

a. Both the SN and destination nodes compute the source-destination shared session key $\phi_{sn \rightarrow ds}$ as follows:

i. SN and destination nodes uniquely generate κ_1 and κ_2, respectively.

ii. SN extracts the two-way distribution parameter of destination node β_{ds} to compute $\phi_{sn \rightarrow ds}$ as $\phi_{sn \rightarrow ds} = \kappa_1 \beta_{ds}$.

b. Sign its data M using its source-destination shared session key $\phi_{sn \rightarrow ds}$ as $\delta = H\, \phi_{sn \rightarrow ds}\, (M)$, perturbs M as $P_p = M + \lambda_n$.

c. SN finally generates its message packet as $\sigma = \delta || P_p || F_i || F_j || n$, and encrypts it as $\sigma_\sigma = \sigma \oplus \phi_{sn \rightarrow ds}$ to further ensure second-tier data confidentiality and integrity of the message and communication information, where F_i and F_j are the pseudonyms of the source and destination nodes, respectively.

d. SN then performs PoP authentication with its IN, as described in the next section, before hopping P_p to the IN.

3.2.5.5 Authentication phase

After the signature and perturbation phase, the source node initiates the PoP authentication with the IN as follows:

i. SN generates an authentication token ω and time stamp t_s.

ii. SN and IN randomly generate $\upsilon \in Z_q^*$ and $\varepsilon \in Z_q^*$, respectively. SN computes its PoP authentication parameter as $n_{sn} = (\upsilon + \mu)P \bmod q$, while IN computes its own as $n_{in} = (\varepsilon + \mu)P \bmod q$ and sends it to SN, who then computes its PoP session authentication key $\varphi_{sn \rightarrow in}$ as $\varphi_{sn \rightarrow in} = \upsilon . n_{in}$.

iii. SN then encrypts the concatenated authentication token ω, pseudonym of source, pseudonym of IN, and time stamp as $E\varphi_{sn \rightarrow in}\, (\omega || F_{sn} || F_{in} || t_s)$, concatenates it with n_{sn} as $E\varphi_{sn \rightarrow in}(\omega || F_{sn} || F_{in} || t_s)) || n_{sn}$, and sends it to its IN.

iv. On the receipt of $E\varphi_{sn \rightarrow in}(\omega\, || F_{sn} || F_{in} || t_s) || n_{sn}$, IN extracts n_{sn} then computes its $\varphi_{\,in \rightarrow sn} = \varepsilon . n_{sn}$. It decrypts the received $E\varphi_{sn \rightarrow in}(\omega || F_{sn} || F_{in} || t_s)$ using its $\varphi_{sn \rightarrow in}$ to extract ω and t_s. It, thereafter, re-encrypts the extracted ω and t_s, using $\varphi_{in \rightarrow sn}$, and sends it back to the SN. The SN decrypts it using its $\varphi_{sn \rightarrow in}$ and verifies it by comparing the ω and t_s with their original values. If equal, SN hops its encrypted data packet σ_σ. The IN then becomes temporary SN and repeats this phase with its selected IN until the packet gets to the destination node.t

3.2.5.6 Verification and decryption

Destination node extracts and authenticates the received data M by following this procedure:

i. Destination node extracts the two-way distribution parameter of SN and βsn and computes destination of the used perturb P.

ii. Destination node regenerates the used perturb λ'_n by checking the value on the perturb index n. If $n = 1$, it indicates that the source is new to the destination node, and destination node then executes perturbation generation phase in order to obtain the perturb seed, which would be used to recompute the used perturb. However, if $n > 1$, it indicates that the session is for old destination node. The destination node retrieves the encrypted last perturb for the source node from its memory, decrypts it, and uses it to obtain the used perturb by executing step 3 of the perturbation generation phase. Extract the n message by unperturb P_P as: $M' = P_P - \lambda'_n$.

iii. Destination node verifies the signature by re-signing the unblinded message M' using its $\phi_{ds \to sn}$ as $\delta' = H\phi_{ds \to sn}(M')$. If $\delta' = \delta$, then the perturb, data, and the source node are all valid, and destination node then accepts the data, otherwise rejects the data. Encrypt the perturbation parameter as $\lambda_n \oplus \vartheta$, $m_2 \oplus \vartheta$, $n \oplus \vartheta$, F_{sn}. Clear all the previously encrypted perturbation parameters stored for F_{sn} in its memory, and replace it $(\lambda n||\vartheta||m2||\vartheta||n)$.

4. Conclusion

This chapter shows overview of wireless sensor networks with its security and privacy framework. The chapter proffers to readers an in-depth understanding of security and privacy issues as related to WSNs. Some existing research in WSN routing protocols are discussed. This chapter also helps researchers to understand the current trends in WSNs routing protocols and security schemes.

Author details

Oladayo Olufemi Olakanmi* and Adedamola Dada
Department of Electrical and Electronic Engineering, University of Ibadan, Nigeria

*Address all correspondence to: olakanmi@mit.edu

References

[1] Oladayo O, Abass A. A secure and energy-aware routing protocol for optimal routing in mobile wireless sensor networks (MWSNs). International Journal of Sensors, Wireless Communications and Control. 2019;**9**(Pt 4)

[2] Sohrabi K, Gao J, Ailawadhi V, Pottie GJ. Protocols for self-organization of a wireless sensor network. IEEE Personal Communications. 2000;7 (Pt 5):16-27

[3] Villalba LJG, Orozco ALS, Cabrera AT, Abbas CJB. Routing protocols in wireless sensor networks. International Journal of Medical Sciences. 2009:8399-8421

[4] Messaoudi A, Elkamel R, Helali A, Bouallegue R. Cross-layer based routing protocol for wireless sensor networks using a fuzzy logic module. In: Paper Presented at the 13th International Wireless Communications and Mobile Computing Conference (IWCMC); 2017

[5] Huei-Wen DR. A secure routing protocol for wireless sensor networks with consideration of energy efficiency. In: IEEE National Taiwan University of Science and Technology; 2012. pp. 224-32

[6] Murugaboopathi G, Khaana V. Reliable communications in sensor networks. Journal of Engineering and Applied Science. 2008;**3**:911-917. Available from: https://medwelljournals.com/abstract/?doi=jeasci.2008.911.917 [Accessed: 10 July 2017]

[7] Saraswati M, Prabhjot K. Energy efficient neighbor selection for flat wireless sensor networks. Information Technology and Management. 2013: 518-523

[8] Lv S, Wang X, Zhao X, Zhou X. Detecting the Sybil attack cooperatively in wireless sensor networks. In: Paper Presented at the International Conference on Computational Intelligence and Security; 2008

[9] Olakanmi O, Dada A. An efficient point-to-point security solution for multi-hop routing in wireless sensor networks. Security and Privacy. 2018

[10] Oladayo O, Adama P. An efficient multipath routing protocol for decentralized wireless sensor networks for mission and safety-critical systems. International Journal of Sensors, Wireless Communications and Control. 2019;**9**(Pt 4)

[11] Jamil I, Imad M. A secure hierarchical routing protocol for wireless sensor networks. In: Paper Presented on the 10th IEEE International Conference on Communication Systems; Singapore; 2006

[12] Du X, Xiao Y, Chen H-H, Wu Q. Secure cell relay routing protocol for sensor networks. Special Issue on Network Security. 2009;**6**(Pt 3): 375-391

[13] Masruroh SU, Sabran KU. Emergency-aware and QoS based routing protocol in wireless sensor network. In: Paper Presented at the IEEE International Conference on Intelligent Autonomous Agents, Network and Systems; 2014

[14] Manjeshwar A, Agrawal DP. TEEN: A routing protocol for enhanced efficiency in wireless sensor networks. In: Proceedings of 15th International Parallel and Distributed Processing Symposium; IPDPS; 2009-2015; San Francisco; 2001

[15] Patil M, Biradar RC. A survey on routing protocols in wireless sensor networks. In: Paper Presented at the

18th IEEE International Conference on Networks; Mandya; 2012

[16] Durrani N, Kafi N, Shamsi J, Haider W, Abbsi A. Secure multi-hop routing protocols in wireless sensor networks: Requirements, challenges, and solutions. In: Paper Presented at the 8th IEEE International Conference on Digital Information Management (ICDIM); 2013

[17] Goyal D, Tripathy MR. Routing protocol in wireless sensor networks: A survey. In: Paper Presented at the IEEE International Conference on Advanced Computing and Communication Technologies; Haryana; January 2012

[18] Lu Y, Wang G, Jia W, Peng S. Multipath-based segment-by-segment routing protocol in MANETs. In: Paper Presented at the 9th International Conference for Young Computer Scientists; Hunan; November 2008

[19] Shruti UK. Few locations based routing protocols in wireless sensor network. In: Paper Presented at International Conference on Green Computing and Internet of Things (ICGCloT); Chennai; 2015

[20] Amjad M, Afzal MK, Umer T, Kim B-S. QoS-aware and heterogeneously clustered routing protocol for wireless sensor networks. IEEE Access. 2017;**5**:10250-10262

[21] Raghunandan GH, Lakshmi BN. A comparative analysis of routing techniques for wireless sensor networks. In: Paper Presented at the IEEE Conference on Innovations in Emerging Technology; 2011. pp. 17-22

[22] Krishna KK, Augustine R. A survey on mobility based routing protocols in wireless sensor networks. International Journal of Computer Applications. 2016; **135**(5):36-38

[23] Kim B-S, Kang SY, Lim JH, Kim KH, Krishna KK, Augustine R. A survey on mobility based routing protocols in wireless sensor networks. International Journal of Computers. 2017

[24] Kumar J, Tripathi S, Tiwari RK. Routing protocol for wireless sensor networks using swarm intelligence-ACO with ECPSOA. In: Paper Presented at the International Conference of Information Technology; 2016

[25] Obaidat MS, Li J-S. Security in wireless sensor networks. Security and Communication Networks. 2016;**1** (Pt 1):101-103

[26] Zhang J, Varadharajan V. A new security scheme for wireless sensor networks. In: IEEE Global Communications Conference (GLOBECOM) Proceedings; Hong Kong; 2008. pp. 1-5

[27] Qi J, Hong T, Xiaohui K, Qiang L. Detection and defence of sinkhole attack in wireless sensor network. In: Paper Presented at the IEEE 14th International Conference on Communication Technology; Chengdu; 2012

[28] Amisha P, Vaghelab VB. Detection and prevention of wormhole attack in wireless sensor network using AOMDV protocol. In: Paper Presented at the 7th International Conference on Communication, Computing and Virtualization; 2016

[29] Ahmad Salehi S, Razzaque MA, Naraei P, Farrokhtala A. Detection of sinkhole attack in wireless sensor networks. In: Paper Presented at the IEEE International Conference on Space Science and Communication (IconSpace); 2013

[30] Magotra S, Kumar K. Detection of HELLO flood attack on LEACH protocol. In: Paper Presented at the IEEE

International Advance Computing Conference (IACC); 2015

[31] Xiangning F, Yulin S. Improvement on LEACH protocol of wireless sensor network. In: Paper Presented at the IEEE International Conference on Sensor Technologies and Applications; 2007

[32] Raymond DR, Midkiff SF. Denial-of-service in wireless sensor networks: Attacks and defences. IEEE Pervasive Computing. 2008;**7**(Pt 1):74-81

[33] Ramkumar M. Proxy aided key pre-distribution schemes for sensor networks. In: Paper Presented at the IEEE International Conference on Performance, Computing, and Communications; 2010. pp. 461-68

[34] Gungor VC, Lu B, Hancke GP. Opportunities and challenges of wireless sensor networks in smart grid. IEEE Transactions on Industrial Electronics. 2010;**57**:3557-3564

[35] Du X, Guizani M, Xiao Y, Chen H-H. An effective key management scheme for heterogeneous sensor networks Francisco; December 2006

[36] Yu Z. The scheme of public key infrastructure for improving wireless sensor networks security. In: Paper Presented at the IEEE International Conference on Computer Science and Automation Engineering IEEE Transactions; Manchester; 2012

[37] Chen J-L, Lai Y-F, Lu H-F, Kuo Q-C. Public-key based security scheme for wireless sensor network. In: IEEE Radio and Wireless Symposium 2008. pp. 255-258

[38] Du X. Two tier secure routing protocol for heterogeneous sensor networks. IEEE Transactions on Wireless Communications. 2007; **6**(Pt 9):3395-3401

[39] Du X. QoS routing based on multi-class nodes for mobile. Ad Hoc Networks. 2004;**2**(Pt 3):241-254

[40] Abdul Hamid M, Mustazur Rahman M, Yoon YJ, Hong CS. Developing a group-based security scheme for wireless sensor networks. In: IEEE Global Communications Conference (GLOBECOM) Proceedings; 2007

[41] Li-Ping Z, Yi W, Gui-Ling L. A novel group key agreement protocol for wireless sensor networks. In: Paper Presented at the International Conference on Wireless Communication and Signal Processing; Beijing; 2009

Chapter 3

Design of an Ad Hoc Mesh Network for Aircrafts

Ichi Kanaya and Eri Itoh

Abstract

This article reports an exploration of the information flow among aircrafts and describes a novel digital communication protocol which uses ad hoc mesh networking technologies. The proposed process may be operated utilizing current aircraft hardware and accomplishes very dependable interaction with a quick time period (a couple of tens of seconds). Simulations check that over 200 [octet] of information is usually shared with 98% of the aircraft inside a chosen region.

Keywords: wireless communication, ad hoc network, mesh network, internet protocol, air traffic management

1. Introduction

Communications among aircrafts (air to air) and between an aircraft and the ground control (air to ground) play a crucial part in the secure operation of the aviation business. Flight Deck Interval Management (FIM) that exchanges place as well as altitude information straight between planes is a major engineering for Continuous Descent Operation (CDO), resulting in extremely effective aircraft operations. Underneath CDO, arrival plane descends at a continuous velocity to the runway at a state close to idling; current study results have shown the power effectiveness of this particular strategy [1, 2].

Conventional analysis suggests that the Aircraft Surveillance Applications System (ASAS) may well not have the ability to deal with the projected intense increase in the variety of the plane until the authors are able to conquer the bottleneck of information sharing involving aircrafts. ASAS enables the surveillance of adjacent traffic flow in the environment with the traditional Automatic Dependent Surveillance Broadcast (ADS-B), which are only able to process incredibly limited quantities of info, which means that innovative correspondence is not possible [3].

Even though satellite and the ground-based internet for aircraft usually do not address whole routes, they offer non-mission-critical internet services to passengers. Additionally, they usually need the assembly of a new antenna on the plane.

In this analysis, the authors think about the setup of an ad hoc mesh system which employs the current aircraft communication tools almost as practical and will discuss adequate information for the FIM. This proposal is dependent on the concept that the existing aircraft facilities are untouched, making it possible for a system to be recognized without renovating the outside of the plane. Additionally, the proposed technique is powerful against unexpected accidents including equipment breakdown, since it does not need a predesigned topology and can transmit data redundantly throughout the mesh network.

This concept is complementary to the standard internet technology for aircraft like satellite web. It's potential to run web-based FIM even once the plane is outside of the coverage of internet services.

2. Related work

The authors discuss about connected researches on candidates for inter-aircraft computer networks in terms of the Open Systems Interconnection (OSI) reference model. The authors likewise outline the current radio equipment found in a regular aircraft.

2.1 Physical and datalink layers

The authors discuss IEEE 802.11 (Wi-Fi), IEEE 802.16 m (WiMAX), and IEEE 802.15.4 and IEEE 802.22 that are standardized by the IEEE 802 standardization committee for local area networks (LAN) and metropolitan area networks (MAN). The authors subsequently present an overview of satellite internet as well as the ground-based internet. Lastly, the authors are going to touch on the communication technologies presently used in the plane.

IEEE 802.11a/g/n/ac (also known as Wi-Fi) is a physical and datalink layer protocol that uses 2.4–2.5, 4.9–5.0, 5.03–5.091, 5.15–5.35, and 5.47–5.725 GHz. The modulation is performed by orthogonal frequency-division multiplexing (OFDM) that uses orthogonal subcarriers [4–6].

IEEE 802.16 m (also known as WiMAX) is a physical/datalink layer protocol that uses wireless networking for communication over a wider space than the IEEE 802.11a/g/n/ac series. It often uses 2.575–2.645 GHz UHF band and OFDM [7, 8].

Long-Term Evolution (LTE) is a standard for mobile phones. In Japan roughly 0.7–1.7 GHz (UHF band) is employed. It uses quadrature phase-shift keying (QPSK) over OFDM in every subcarrier [9–11].

IEEE 802.15.4 and IEEE 802.22 are less famous than the above three protocols. IEEE 802.15.4 is a physical/datalink layer protocol for short-range wireless communication which uses 868.0–868.8, 902–928 MHz, or 2.4000–2.4835 GHz and direct-sequence spread spectrum (DSSS) or offset quadrature phase-shift keying (OQPSK). IEEE 802.15.4 is best widely known as Zigbee wireless network protocol. IEEE 802.22 is a standard for wireless regional space networks using white areas within the TV broadcast bands [12].

Satellite net services are primarily provided by geostationary satellites, although some are provided by satellite constellations who relay communications via many satellites. As fixed satellites do not handle the polar regions, therefore the communication coverage of theirs is restricted.

Since satellite constellations are able to work with low-earth-orbit satellites, the correspondence latency is smaller than within the net service provided by fixed satellites. Because the places of those satellites are completely dynamical, an omnidirectional antenna is usually required, and far additional power is required than for communication with fixed satellites [13, 14].

The authors here shortly discuss about ground-based aircraft internet, aircraft-based internet, and latest developments in long-distance and low-speed network engineering for the Internet of Things (IoT).

References [15–25] show other popular wide-area network technologies.

A voice communication channel using 118.0–136.975 MHz (VHF) is also used. The audio uses amplitude modulation (AM).

2.2 Network layer

In the network layer protocol, the Internet Protocol (IP) is the factual standard whenever the datalink level has enough bandwidth. Based on the IP, the Space Communications Protocol Specifications (SCPS) may also be used when the band is extremely narrow, like for communication between the earth and other planets [26, 27].

2.3 Transport layer and above

It's typical for the transport layer to ensure end-to-end communication. In networks based on the IP for the network layer, the Transmission Control Protocol (TCP) is often used for the transport layer [28]. Networks which do not make use of IP for the network layer, or networks which use IP but do not use TCP for the transport layer, occasionally have the own error-correction mechanism of theirs in the transport layer [29, 30].

The authors do not go over the session layer (or higher layers), as they are not the topic of the research.

3. Design of networking protocol

In the prior section, the authors summarized the conventional wireless network technologies. They're, nonetheless, unsatisfactory due to the constraints of the conventional aircrafts.

IEEE 802.11, IEEE 802.15.4, IEEE 802.16 m, and LTE will need the aircrafts to be engineered with freshly installed antennas and are probably jammed by high-speed movement and may experience radio interference since they are acting secondary modulation by OFDM.

As satellite internet utilizing geostationary satellites does not handle the polar regions, it cannot be put on to aircraft traveling near the North Pole. Furthermore, satellite internet utilizing satellite constellations necessitates broadcasting communication among the satellites, which is technically complicated.

Air-based internet and ground-based internet aren't ideal for FIM because the coverage area of theirs is limited.

Long-distance, low-speed networks (excluding LoRa) aren't ideal for FIM, because interaction with mobile vehicles is not taken into consideration in the design of theirs.

Consequently, the authors have created a brand new low-speed (1–10 kbps) wireless communication technology to allow the inter-aircraft flow of information. The authors assume that conventional wireless communications are used by the aircraft in the physical layer and the datalink layer for the best backward compatibility with existing aircrafts.

The authors present the following communication protocol and then verify the effectiveness of the protocol by simulations.

3.1 Physical layer

The authors assume narrowband wireless communication in the physical layer. In order to share the antenna with SSR [31, 32] the authors utilize the UHF band as well as frequency modulation (FM) that is ideal for digital signals, and they have high noise tolerance.

In thought of the practicableness of implementing this specific protocol on typical aircraft, the authors use precisely the same band as SSR (i.e., the UHF band) to enable the potential for diverting the SSR antenna.

As the current SSR uses a straightforward pulse-based communication protocol, the communication band is incredibly narrow. Thus, a bandwidth of roughly 10 kHz with adjacent frequency as being a carrier is available. The authors consider FM for the noise tolerance at this layer. Generally, AM is utilized for the lower selectivity of voice communication channels in aircraft communication, but as this proposal is restricted to digital communication, FM is adopted since it is more reluctant to interference compared to AM.

3.2 Datalink layer

The datalink layer encodes/decodes the data we send/receive with metadata including data sender, time code, etc. These packed data are referred to as the packets.

Permit the size of the packet at the season of exchange from the datalink layer to the physical layer be near 256 [octet]. As the littlest metadata, the authors incorporate a "magic" code up to 4 [octet] specifying the character of the packet, a time code (4 [octet]), and a code of the aircraft (4 [octet]). By considering these headers, 240 [octet] is left inside the datagram. Expecting that 2 kbps is offered as the channel of the datalink layer, it will take 1 s to transmit a solitary packet. The transmission capacity of 2 kbps could be feasible.

In detail, aircraft code (4 [octet]) contains the source aircraft code. A unique number for every aircraft must be allocated. The body (240 [octet]), followed by error-detection code, contains data body. The error-detection code (4 [octet]) contains error detection. CRC-32, which inspects the cyclic redundancy check, is used.

Packets having errors and packets older than a planned threshold are discarded. To scale back the load on the network layer, it manages "time to live" of the packet.

3.3 Network layer

The network layer encodes/decodes the sent/received data by wrapping/unwrapping the datalink packet. Dropping any irregular data is done in the network layer. If there are not any irregularities, the data is passed to the upper layer, that is, the transport layer. If needed, the received data is retransmitted. A signal body of the transport layer is named a datagram.

The datagram includes a code (up to 4 [octet]) stating the character of the datagram, a destination aircraft code (4 [octet]), a source aircraft code (4 [octet]), and a time code (4 [octet]). The size of the user space of the datagram is 224 [octet].

3.4 Transport layer

In contrast to the internet, it's really hard to assure end-to-end data transmission in aircraft communication. Therefore, the transport layer does not handle the end-to-end connection feature, that's characteristic of TCP, like data retransmission requests. Yet, a comparatively powerful error-correction capability is enforced to the transport layer. The Reed-Solomon (RS) code is thought to be the promising candidate. When three datagrams are transmitted together with error-correction signals distributed in four datagrams, the typical quantity of data per datagram is 168 [octet].

The Reed-Solomon code, which is a more practical error-correction function than the cyclic redundancy check (CRC), is enforced within the transport layer. If three datagrams are coupled with the error-correction signal and, as a result, distributed to four datagrams to be transmitted, 168 [octets] per datagram are allotted to the implementer.

4. Feasibility study of physical, datalink, and network layers

In order to confirm the feasibility of the protocol discussed in the prior section, the authors developed the model described below and conducted a simulation experiment.

Assume that N aircrafts a_i are existing at positions p_i, respectively. Position p_i may or may not be a three-dimensional orthonormal coordinate system. The primary point in this discussion is that the distance $p_i - p_j \vee$ between position p_i and position p_j is defined.

Let x_{ij} be the information the authors want to send to aircraft a_j from aircraft a_i. The information received by aircraft a_i is, nonetheless, different from x_{ij}—the authors denote it as y_{ij}.

When there's no error in the communication route, the information x_{ij} originating from aircraft a_i corresponds to y_j received by aircraft a_j. This can be described as

$$y_j = Ui\, x_{ij} \tag{1}$$

Let us denote the transmission rate of the communication route from aircraft a_i to aircraft a_j by c_{ij}. The transmission rate c_{ij} refers to the packet data transmitted by aircraft a_i that is actually received by aircraft a_j, that is, the probability of correct information transmission. Let c_{ij} be a function of position p_i and position p_j. That is, $c_{ij} = e(p_i, p_j)$. In wireless communication, we can reasonably assume that

$$c_{ij} = k\,\varepsilon^{-K|p_i - p_j|^2} \tag{2}$$

for some constants k, K. These parameters may be changed according to the results of experiments. In this report, the authors adopt Eq. (2) as the transmission rate c_{ij}.

Taking the transmission rate into account, Eq. (1) is modified to

$$y_j = Ui\, c_{ij} x_{ij} \tag{3}$$

Until now, the authors have dealt with only one-to-one communication from aircraft a_i to a_j, but we can consider another aircraft a_k relaying the communication described by x_{ij}. When the information that aircraft a_k receives and retransmits is z_k, where $z_k = ij\, c_{ik} x_{ij}$, there would be an explosive increase in the amount of data if the authors do not use an artificial attenuation (decay) term d_{ijk}, as in

$$z_k = Uij\, c_{ik} d_{ij} k\, x_{ij}. \tag{4}$$

Here, the attenuation term d_{ijk} denotes the probability of intentionally discarding the packet during the packet relay. Finally, the equation that takes the relay into consideration has the form

$$y_j = Ui\, c_{ij} (x_{ij} \cup z_{ij}). \tag{5}$$

In this report, to statistically investigate the arrival rate of data based on Eqs. (4) and (5), the authors performed computer simulation with the following parameters:

Airspace. Three-dimensional orthonormal space. It is a cube whose vertexes are (0, 0, 0)-(0, 0, 1)-(0, 1, 0)-(0, 1, 1)-(1, 0, 0)-(1, 0, 1)-(1, 1, 0)-(1, 1, 1). The units are arbitrary.

Time. The simulation is performed between 0 and 100 s with a time step of 1 s.

Number of aircraft. 1000 aircraft are randomly placed in the airspace.

Probability of successful communication. The probability of the successful communication is based on Eq. (2) according to the interval between the aircrafts. However, the authors set $k = 1$ and simulate K as $K \in \{0.1, 0.316, 1\}$. **Figure 1** shows the probability of success depending on the interval for each parameter.

For every simulation, the authors transmitted data from aircraft a_{999} to a_0 in each time step. If the data did not reach a_0 in a single time step, and if they reached a_n where $n \neq 0$, then a_n attempted to send the data to a_0 in the next time step.

Theoretically, the attenuation term d_{ijk} in Eq. (4) must be equal to or less than 1, and the authors assumed that $d_{ijk} = 1$ in the simulations to clarify the experimental results.

The authors have done the following two investigations.

Simulation 1. Investigate the degree of data transmission per time step using multi-hop communication via neighboring aircraft wherever the communication path is unstable.

Simulation 2. Investigate the entire range of hops for every time step using multi-hop communication via neighboring aircraft wherever the communication path is unstable.

5. Experimental results

Figure 1 shows results of Simulation 1. **Figures 2** and **3** show the results of Simulation 2.

In **Figure 2**, the horizontal axis represents time steps, and the vertical axis represents the amount of aircraft (maximum 1000). The blue curve shows the case of $K = 0.1$, the green curve shows the case of $K = 0.316$, and the yellow curve shows the case of $K = 1$. All curves were plotted under the condition of $k = 1$.

Typical values of **Figure 2** are picked up in **Table 1**. Time steps 1, 30, and 60 correspond to 1, 30, and 60 s after the start of the communication, respectively.

The horizontal axis and the vertical axis of **Figure 3** represent generations and the overall range of hops needed for data transmission, respectively. The blue curve shows the case of $K = 0.1$, the green curve shows the case of $K = 0.316$, and the yellow shows the case of $K = 1$ under the condition of $k = 1$ in Eq. (2).

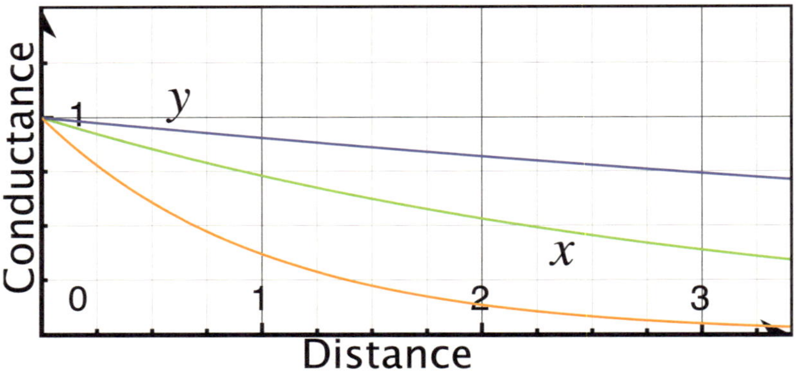

Figure 1.
Simulation parameters (yellow, K = 0.1; green, K = 0.316; blue, K = 1).

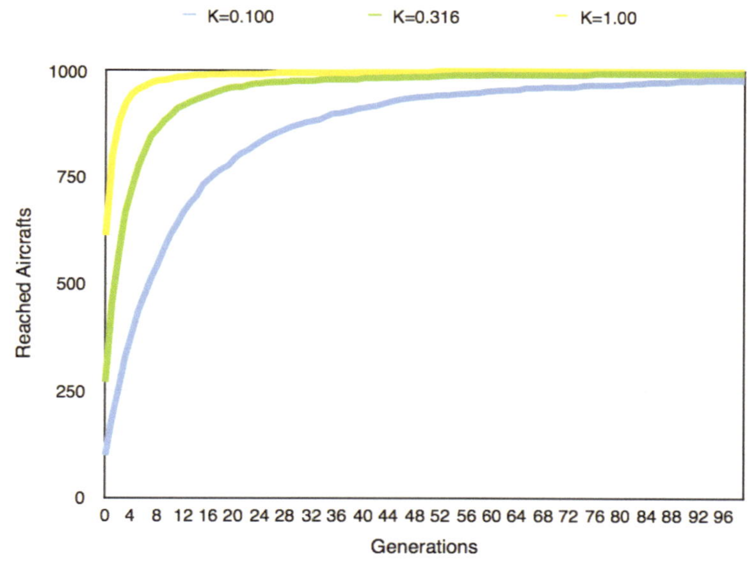

Figure 2.
Number of aircraft reached.

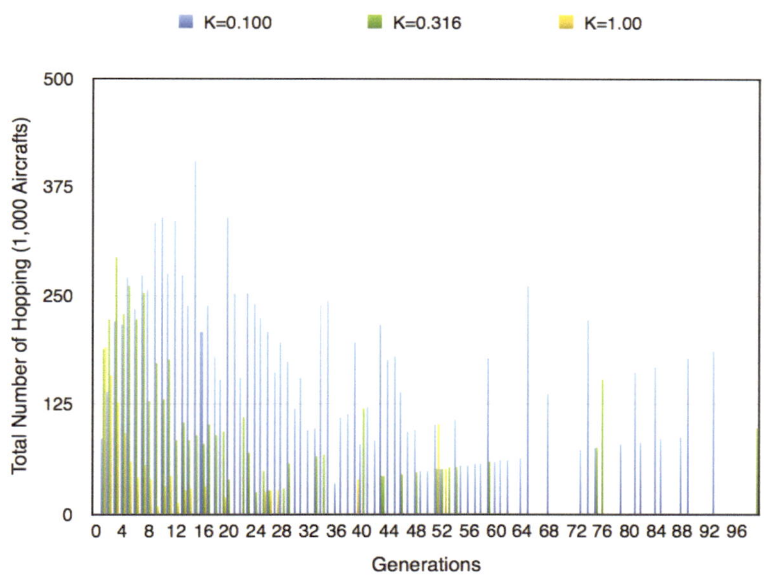

Figure 3.
Number of hops.

K	Generation 1	Generation 30	Generation 60
0.1	98	871	953
0.316	269	977	992
1	612	996	1000

Table 1.
Velocity of the spread of shared information.

6. Discussion

Simulation 1 under the condition of $K = 0.316$ shows that 97.7% of information is effectively transmitted by time step 30 (which means 30 s after the start of the communication). If we remind $\varepsilon^{-0.316} = 0.729$, we can say that the theoretical arrival rate of information with a distance of one in this model is 72.9%, while by this method, the rate can be improved to 97.7% (meaning 134% improvement) within 30 s, and this improvement can be achieved without controlling intervals of the aircrafts.

By the outcomes of Simulation 2, the total amount of hops necessary for information transmission is likely to decrease with a rise in the quantity of time steps. In this particular operation, its anticipated attenuation term d_{ijk} is set to less than 1, which means that attenuation is intentionally performed.

As discussed above, even in a communication environment with extremely small reliability, it's apparent that highly dependable communication is possible through the dispersed nature of the aircrafts in the airspace. Under the circumstances of this particular simulation, 97.7% of information was correctly shared among the aircrafts with 30 s of interaction. As the datagram size is no more than 224 [octet], aside from error correction in the transport layer, the data rate under a guaranteed transmission rate of 98% is approximately 60 bit/s in the worst case. (In the very best situation, the theoretical value is approximately 1.8 kbit/s).

Even though this appears to be an incredibly narrow communication band for contemporary wireless communication, it's a feasible numerical value for mission-critical inter-aircraft. For communication which is not mission critical, we are able to consider satellite internet as a complementary protocol.

7. Conclusion

In this article, the authors reported an investigation of the information flow among aircrafts and proposed a new digital communication protocol which uses ad hoc mesh network technologies. The authors illustrated the proposed protocol could be operated utilizing conventional aircraft hardware and will accomplish extremely dependable interaction with a very short time period (a couple of tens of seconds). The simulations supported the strength of the suggested protocol.

If the ad hoc mesh system introduced in this article will come into reality, the application will cover not only CDO but also the autonomous management of associate adjacent craft. Moreover, computer networks that will manage ultrahigh-speed nodes like aircraft would contribute to conventional mobile digital networks that are actively studied.

Acknowledgements

This research was supported by *Electronic Navigation Research Institute*, Japan.

A. Simulation program

The simulation program used in this report is posted at https://github.com/kanaya. This program works with the Scheme interpreter (R5RS conforming or higher).

Author details

Ichi Kanaya[1]* and Eri Itoh[2]

1 The University of Nagasaki, Nagasaki, Japan

2 Electronic Navigation Research Institute, Tokyo, Japan

*Address all correspondence to: kanaya@pineapple.cc

References

[1] Kanaya I. Proposal of inter-aircraft mesh computer network. Research Report. Japan: Electronic Navigation Research Institute, ENRI; 2016

[2] Itoh E, Brown M, Wickramasinghe N, Fukushima S. Future arrival management collaborating with trajectory-based operations. In: Air Traffic Management and Systems II. Germany: Springer; 2017

[3] Richards WR, O'Brien K, Miller DC. New air Traf surveillance technology. Aero Boeing. 2010;**02**:7-14

[4] Reid NP, Seide R. Wi-Fi (802.11) Network Handbook. USA: Osborne Networking; 2002. ISBN: 978-0072226232

[5] LitePoint. IEEE 802.11ac: What Does it Mean for Test? LitePoint Whitepaper. 2013

[6] IEEE Standard Association. IEEE Standard for local and metropolitan area networks Part 15.4: Low-Rate Wireless Personal Area Networks (LP-WPANs). IEEE Computer Society; 2011

[7] Fazel K, Kaiser S. Multi-Carrier and Spread Spectrum Systems: From OFDM and MC-CDMA to LTE and WiMAX. 2nd ed. USA: John Wiley & Sons; 2008. ISBN: 978-0-470-99821-2

[8] Richardson M, Ryan PS. WiMAX: Opportunity or Hype? ITERA; 2006. Available from: https://ssrn.com/abstract=892260

[9] Sesia S, Baker M, Toufik I. LTE: The UMTS Long Term Evolution: From Theory to Practice. USA: Wiley; 2011

[10] Dahlman E, Ekstrom H, Furuskar A, Karlsson J, Meyer M, Parkvall S, et al. The long-term evolution of 3G. Ericsson Review, No. 02; 2005

[11] Alcatel-Lucent. Using air-to-ground LTE for in-flight ultra-broadband. Alcatel-Lucent Strategic White Paper. Alcatel-Lucent; 2015

[12] Mody AN. IEEE 802.22 Wireless Regional Area Networks—Enabling Rural Broadband Wireless Access Using Cognitive Radio Technology. IEEE; 2010

[13] Walker JG. Satellite constellations. Journal of the British Interplanetary Society. 1984;**37**:559-571

[14] Ballard AH. Rosette constellations of earth satellites. IEEE Transactions on Aerospace and Electronic Systems. 1980;**16**(5):656-673

[15] Stotts LB, Plasson N, Martin TW. Progress towards reliable free-space optical networks. In: Military Communications Conference (MILCOM) 2011; IEEE; 2011. DOI: 10.1109/MILCOM.2011.6127559

[16] Gogo. Gogo ATG-4: What is it, and how does it work? Gogo Press Release; 2014. Available from: https://concourse.gogoair.com/gogo-atg-4-work/ [Accessed: 25 July 2017]

[17] Thompson P. How In-Flight WiFi Works And Why It Should Get Better; Jalopnik; Issue of 20/6/14; 2014. Available from: https://bit.ly/2eJpBy6 [Accessed: 25 July 2017]

[18] Anthony S. DARPA begins work on 100Gbps wireless tech with 120-mile range; Extreme Tech. Issue of 17/12/2012. Available from: https://bit.ly/2uUijh8 [Accessed: 25 July 2017]

[19] Miura R, Adachi F, Tada M, Yonemoto N, Watanabe S. R&D on Cooperative Technologies between Unmanned Aircraft Systems (UAS)-Based Wireless Relay Systems and

Terrestrial Networks with Frequency Sharing; System of Radio Use. Japan: Ministry of Internal Affairs and Communications; 2016

[20] Metz C. Inside Facebook's First Efforts to Rain Internet from The Sky. Wired; 2016

[21] Massemin E. Internet des objets, Ludovic Le Moan (Sigfox) lance l'IoT Valley a Labage; La Tribune Toulouse; Issue of 22/05/2012

[22] Kalfus R, Hegr T. Ultra narrow band radio technology in high-density built-up areas. In: Information and Software Technologies: 22nd Int'l Conf., ICIST 2016; October 13-15, 2016; Druskininkai, Lithuania: Proc., Springer; 2016

[23] EETimes Japan. Sony presented LPWA which communicated over 100km [in Japanese], EETimes Japan; Issue of 9/6/2017. Available from: https://bit.ly/2vDjjnI [Accessed: 25 July 2017]

[24] Prajzler V. LORA, LORAWAN AND LORIOT, LORIOT AG; 2015. Available from: https://bit.ly/2tV6KCp [Accessed: 25 July 2017]

[25] Svetlana G. 3GPP Low power wide area technologies. GSMA White Paper. GSMA; 2016. pp. 49

[26] Siyan KS. Inside TCP/IP. 3rd ed. USA: ACM; 1997. ISBN: 1562057146

[27] Gislason D. Zigbee Wireless Networking. Netherlands: Elsevier; 2008. ISBN: 9780080558622

[28] Kevin R, Fall W, Stevens R. TCP/IP Illustrated, Vol. 1. The Protocols. 2nd ed. USA: Addison-Wesley; 2014. ISBN: 978-9332535954

[29] Internet Engineering Task Force: RFC768. Available from: https://tools.ietf.org/html/rfc768

[30] Internet Engineering Task Force: RFC3550. Available from: https://tools.ietf.org/html/rfc3550

[31] Kayton M, Fried WR. Avionics Navigation Systems. 2nd ed. USA: John Wiley & Sons; 1997. ISBN: 0-471-54795-6

[32] Shipley R. Secondary surveillance radar in ATC systems: A description of the advantages and implications to the controller of the introduction of SSR facilities. Aircraft Engineering and Aerospace Technology. 1971;**43, 1**:20-21

Chapter 4

Key Management Techniques for Wireless Mesh Network

Vinh Truong Quang and Hoa Le Viet

Abstract

Key management is one of the most important tasks in wireless mesh network. This service is responsible for key generation, distribution, and key exchange in a cryptography-based system. Due to the shared nature of WMNs and absence of globally trusted central authority, key management becomes more challenging. This chapter introduces several key management methods that can address these challenges. The fundamental approach is the secret sharing scheme created by A. Shamir, which effectively distributes keys to all participants' network. Based on Shamir's scheme, many authors proposed other algorithms to secure the communication channel in such a way that adversary cannot steal any information about the secret. In addition, in this chapter, a new secret sharing method using real-time synchronization among transceiver devices is presented. In this method, each node generates its key depending on its physical information and the real-time clock. Therefore, public and private keys can be managed efficiently for data encryption and prevent several external attacks to WMNs. A specific protocol is proposed to secure keys while transferring between devices to prevent internal attacks.

Keywords: wireless mesh network, key management, wireless encryption, secret sharing, cryptography

1. Introduction

WMNs are increasingly becoming a prominent architecture that are used in various applications such as home networking, transportation, enterprise networking, etc. [1]. WMNs are very vulnerable to be attacked by opponents. There are three types of attack: active attacks, passive attacks, and message distortion [2]. In order to guarantee the security of data in such networks, cryptography is one of the most popular choices. Therefore, key management services are in demand.

Key management refers to process of cryptographic key generation, distribution, and storage [3]. In addition, the responsibility of key management is establishment of trusted and secure communication between nodes. Due to the unique nature of WMNs, there are three challenges that many existing key management schemes are facing [4]. Firstly, it is difficult to share, transport, and update keys because of the lack of infrastructure in WMNs. Secondly, a distributed certificate authority (CA) model is needed to tackle the absence of fixed central infrastructure in WMNs which is not suitable for public key infrastructure (PKI). Finally, the concern of scalability is undeniable to take advantage of being expandable of WMNs (nodes can join or leave the network).

Many researchers have proposed numerous approaches for group key management. One of the most common group key management methods is secret sharing introduced by Shamir [5]. The schemes allow a master key (secret) to be shared to all authenticated users, but it can just be reconstructed when a node has enough number of shares. Combining with Shamir's method, Li and Xin used the self-certified public key system for proposal of a distributed key management approach [6]. All keys are generated and managed in a self-organizing way within the network, while there is no need of prebuilt trusted relationship between nodes. Lan Yun et al. introduced secret sharing-based management (SSKM) based on Shamir's scheme [7]. The proposed method dynamically generates a different key based on different polynomials from the base station in different periods which can protect the network from the compromised nodes and reduce the high probability of the common key. Filippo Gandino et al. [8] proposed a new key negotiation routine to deal with the case when a node is compromised by adversary. The goal of this algorithm is to reduce the time for the initialization phase as well as reduce the probability of compromised master secret. Singh et al. [9] combined Shamir's scheme and encryption method together by using only hash and XOR function to reduce the overhead for realistic WMNs which have limited resource. All attempts of researchers are to enhance security reliability for key management.

The remaining sections of this chapter are organized as follows. The detail approaches of other authors are introduced in section II, III, and IV. In section V, we proposed a new key management method using real-time synchronization among transceiver devices. In addition, we present our experiments and the result analysis. Finally, the conclusion is drawn in Section VI.

2. Shamir's secret sharing scheme (SSSS)

Shamir's secret sharing scheme (SSSS), also called a (k, n) threshold scheme, is a basic algorithm in cryptography created by Shamir [5]. A secret is divided into multiple parts which each part is given to every participant. To reconstruct the secret, participants have to possess sufficient number of parts. Otherwise, there is no way to reveal the original secret. The goal of Shamir's scheme is to divide S (secret) into n pieces of data $S_1, S_2, \ldots, S_n$ in such a way that knowledge of any k or more S_i pieces makes S easily computable. This means any group of k pieces of data can reconstruct the secret.

Knowledge of $k - 1$ or fewer S_i pieces leaves S completely undetermined. This means secret S cannot be reconstructed with fewer than k pieces.

The (k, n) threshold scheme is based on polynomial interpolation. To build polynomial $f(x) = a_0 + a_1x + a_2x^2 + \ldots + a_{k-1}x^{k-1}$, $(k - 1)$ elements $a_1, a_2, \ldots, a_{k-1}$ are selected randomly and let $a_0 = S$. With a set id_i, $i = 1, \ldots, n$, we calculate $f(id_i)$ and distribute a pair of $(id_i, f(id_i))$ to n participants. For secret recovery phase, we need any subset of k of these pairs. From that, we can find the coefficients of the polynomial using interpolation and evaluate secret $S = f(0)$. To be more efficient, the modular arithmetic is used instead of real arithmetic. The coefficients are randomly chosen in the finite field $GF(q)$. Then, the pairs become $(id_i, f(id_i)) \bmod q)$.

SSSS has a powerful mechanism in key distribution. The number of S_i pieces can be dynamically adjusted according to number of members who join or leave the network. In order to enhance the security level, the shares must be changed frequently without changing the original secret by generating differently the polynomial $f(x)$ with the same free term. SSSS is also suitable for hierarchical system in which the quantity of shares is given to members based on their importance in the system.

The main issue of SSSS comes from cheaters inside the network. In the reconstruction phase, if a member accidently or intentionally provides his fake share, the original secret reconstructed might be wrong as well. Therefore, there is a need of verifying the correctness of the retrieved shares during the reconstruction process.

3. Secret sharing-based key management (SSKM)

Lan Yun et al. [7] presented an algorithm called secret sharing-based key management (SSKM) which can prevent several attacks effectively and reduce the energy consumption. SSKM includes two-level key management. One is to protect communication between base station (BS) and cluster head (CH); another relates to communication between CH and member nodes. Besides, the SSKM utilizes Shamir's secret sharing scheme to distribute keys. The proposed method also dynamically generates different keys based on different polynomials from BS in different periods which can protect the network from the compromised nodes and reduce the high probability of the common keys.

The SSKM was proposed to perform in a hierarchical architecture which consists of a base station and several clusters. Each cluster includes cluster heads and member node. CH manages the cluster and deals with information from member nodes forward to base station (BS).

The SSKM is based on Shamir's scheme to distribute the keys. Though this scheme is information theoretically secured, there are some requirements in this situation:

1. The delivery of shares between dealer and users must operate in a secure channel.

2. The pairs $(x, f(x))$ are made publicly known. However, for security purposes, the pairs must be kept as each user's secret. Therefore, the discrete logarithm in the finite field and DDH (decisional Diffie-Hellman) assumption are adopted to address the problems.

In initial phase, during each session period l $(l = 1,\ldots,M)$, BS randomly generates m polynomials $f(x)$ of $(t-1)$ degree. One of the polynomials $f_{Cin}(x)$ is used to key distribute between BS and cluster heads. Other $m-1$ polynomials are used to key distribute among cluster head and member nodes in $m-1$ clusters, respectively. After that, BS selects M session keys $\{K_{Cin}\}$ and $\{K_{CHi}\}$ from GF(Q) in the finite field Q. $K_{C_{in}}$ is the session key in network key management (between BS and CH), while K_{CH_i} is the session key in cluster key management (between CH and member nodes). Session keys are hidden by calculating $Z_{Cin} = K_{Cin} + S_{Cin}$ and $Z_{CHi} = K_{CHi} + S_{CHi}$; S_{Cin} and S_{CHi} are the secrets. BS/CHs broadcast information Z to each CH/member node.

To protect the communication, the discrete logarithm in the finite field and DDH (decisional Diffie-Hellman) assumption are adopted. As a result, secrets $\{S_{Cin}\}$, $\{S_{CHi}\}$ and session keys $\{K_{Cin}\}$ and $\{K_{CHi}\}$ are kept confidential. After recovering the secret S using (t, n) threshold scheme, users can get session key $K = Z - S$.

SSKM provides an energy-efficient solution in which almost computations were performed by BS, and CHs just exchange parameters to BS to adjust polynomials for key generation/cancelation. SSKM also resolves challenging security issues by localizing key things based on secret sharing scheme. Network key and cluster key management are salient solutions in this work which are mainly responsible for the security protection in a group of members as well as the whole network.

Despite the outstanding advantages, this method is just well-performed in a hierarchical architecture which needs trusted central authorities (BS or CHs). Consequently, the network may be broken down when those authorities are compromised.

4. Hierarchical scheme with transitory master key (HSTMK)

Filippo Gandino et al. [8] introduced a new key management scheme called hierarchical scheme with transitory master key (HSTMK). This approach is based on transitory master key approach and is designed for static wireless sensor network. The proposed scheme includes two elementary key managements. One is plain global key (PKG), which is used by all nodes during the initialization phase and deleted during the working phase. Another is full pairwise keys (FPWK) in which each node shares a specific key with another node. Once an adversary compromises a node before the deletion of key materials, the network is broken down. Therefore, the idea of this method is to split the initialization phase into multi-sub-phase which will increase the overall security level. Furthermore, the HSTMK reduces the time for initialization phase, thus reducing the probability that the master secret is compromised.

A setup task is performed before the initial deployment of the network. In this task, each node is given a common key, called global master key K_{IN}. After that, node u produces its own master key K_U by using a pseudorandom function $f_K(.)$ and global master key K_{IN}:

$$K_U = f_{Kin}(ID_U) \quad (1)$$

Each node has an interval timer which measures the duration of the initialization phase. When the time finishes, all key materials must be deleted to prevent from being stolen by adversaries. This interval value must be selected carefully according to the characteristics of the network.

The initialization phase is divided into four sub-phases, including neighbor discovery, master key computation, pairwise key computation, and acknowledgment.

First, at the neighbor discovery phase, the node broadcast Hello packet to other nodes that identify their neighbors. The packets contain the identification (ID_I) of the senders.

At the master computation phase, the node u calculates the secret of their neighbor v using a pseudorandom function $f_K(.)$ and neighbor's ID_V. Then they delete the global key K_{IN}:

$$K_V = f_{Kin}(ID_V) \quad (2)$$

At the pairwise key computation phase, only one node in a couple of nodes computes the pairwise keys $K_{U\text{-}V}$ and $K_{V\text{-}U}$. Then they delete their neighbor's master key:

$$K_{V-U} = f_{Kv}(ID_u) \text{ or } K_{U-V} = f_{Ku}(ID_V) \quad (3)$$

At the acknowledgment phase, after deleting all key materials, the nodes send acknowledgement messages to authenticate the key establishment.

In addition, the author proposed a mechanism for adding new nodes to network. For the new nodes, at first sub-phase, they broadcast Hello messages to all existing nodes in network. Any nodes which receives these messages will respond with an acknowledgment packet (which contains the ID of the receiver). In the second sub-phase, the new nodes compute the master key of their available neighbors and then delete the global key. Other sub-phases remain unchanged.

An experiment was carried out to compare performance of HSTMK and another method called LEAP+ [10]. The results show that HSTMK is faster than LEAP+ in terms of establishing a pairwise key among nodes. In addition, by increasing the number of nodes in the network, the time before deleting key materials of HSTMK is from 3 to 150 times less than that of LEAPs. The experimental results also highlight the importance of the selection of initialization time. If this value is too low, the proportion of established pairwise keys can reduce, whereas a too long initialization phase would increase the risk of losing secret materials.

5. Real-time key management algorithm

The proposed security algorithm [11] is designed for the purpose of safely transferring keys and synchronous nodes in WMN. In sections 5.1 and 5.2, we will present our key management method based on Adi Shamir's algorithm; the synchronization between nodes by real-time clock helps our keys prevent different types of external attacks. We also present a protocol used for transferring those keys in WMN; this protocol will focus on preventing man-in-middle attack and detecting other abnormal activities in this network.

5.1 Real-time clock key management

The conventional key management methods are easy to be attacked by various attacks such as eavesdropping keys and data, de-authentication, and denial-of-service (DoS). Therefore, we propose to use real-time clock to change continuously private key in key management of each node and synchronize all nodes in WMN, so these nodes will be completely independent of each other. One of those nodes is the network time protocol (NTP) server, and the others are NTP clients. Using the WMN model, the NTP data are transferred quickly enough for synchronization. At a certain point, the nodes will together create a unique key, and every group of n keys is required to reconstruct the same secret for the encryption and decryption.

The process of the proposed method is reversely compared with Adi Shamir's method as shown in **Figure 1**. In the proposed method, the private key is created first instead of the master key. Therefore, the secret will not be detected when the attacker attacks on any node. Besides that, this secret is constantly changed by using a real-time clock module, and thus this makes it more difficult for attackers to be successful in penetrating the network (**Figure 2**).

Private key is generated by a unique value depending on each device—MAC address. A threshold level is required for this process. This parameter will be set depending on the number and installation location of nodes in WMN. The process of private key generation is shown in **Figure 3**.

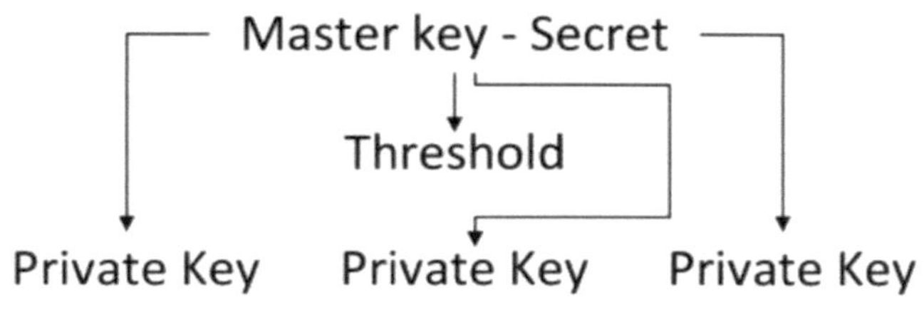

Figure 1.
Adi Shamir's secret sharing scheme.

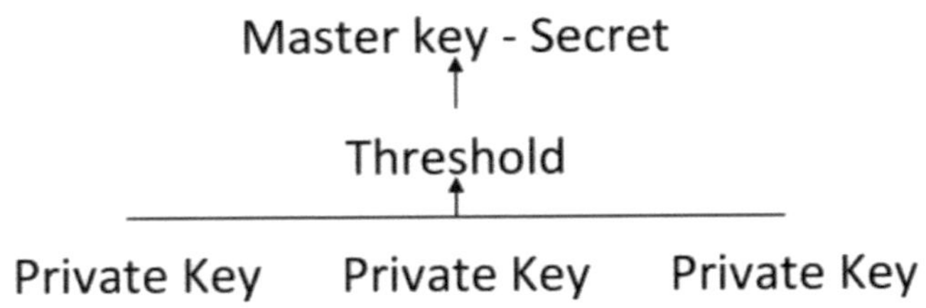

Figure 2.
Proposed secret sharing scheme.

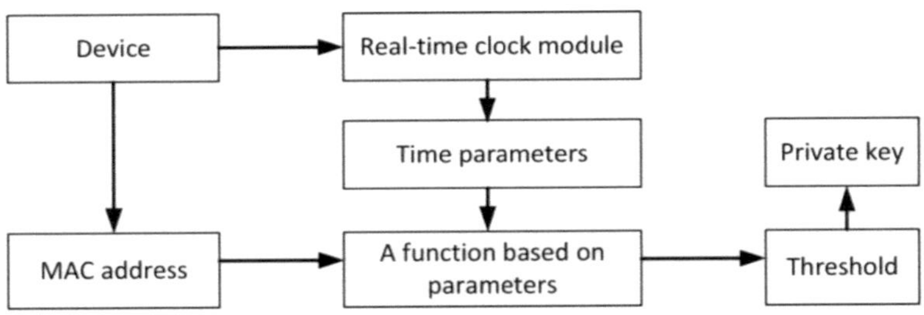

Figure 3.
The process of private key generation.

Reconstructed secret is implemented after each node has enough keys. It is received from nodes in WMN by simple BATMAN protocol which we will talk about in the next section. Lagrange interpolating polynomial was used for our purpose [12]. This is described by the following equations:

For i = (1:node).

$$S = \left(S + f(x_i)\prod_{m=0\wedge m\neq i}^{k-1}(x_m)/(x_m - x_i)\right) \tag{4}$$

If we use this original Lagrange interpolating polynomial, there is a security problem: attackers can gain a lot of information about S with every couple key $(x_i, f(x_i))$. They have numbers to guess from instead of an infinite number of natural numbers by using normal basic methods to solve this set of equations.

This problem can be fixed by using finite field arithmetic in a field of size $p \in P{:}p > S, p > n$. We calculated the couple keys as $(x_i, f(x_i)\ mod\ p)$ instead of $(x_i, f(x_i))$. The lower one sets p, the lower the number of possible values that the attackers have to guess from to set S. Therefore, we made a small change to our key generation function and reconstruction function by the following equations:

$$S = \left(S + p + y.\prod_{k=1\wedge k\neq i}^{k=node}(-x_k).\delta\right) \bmod p \tag{5}$$

$$\delta \times \left(\prod_{k=1\wedge k\neq i}^{k=node}(x_i - x_k)\right) \bmod p \;=\; 1 \tag{6}$$

where S is the secret value which we need for authentication in WMN. Pair of x and y serves as a key to reconstruct secret as we have presented.

5.2 Proposed security protocol

The protocol which we use for our key management scheme is based on the BATMAN protocol—an efficient protocol used to establish connection in WMN. **Figure 4** describes how our protocol works.

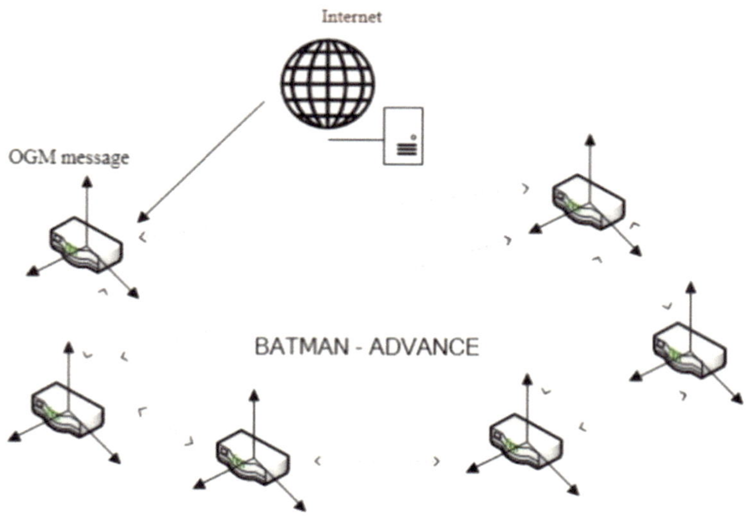

Figure 4.
Modified BATMAN protocol.

While BATMAN advance protocol works as a communicate protocol for sending and receiving data in WMN and detecting node nearby with the same WMN, our proposed protocol uses list neighbor nodes of each node in WMN to work for our purpose. Every node sends its private key frequently to neighbor nodes; an authenticated address list has to be created and checked. This work can be easily done by designing a customized package frame on raw debug socket interface (**Figure 5**).

The objectives of proposed security protocol are the following:

- Encrypt the data by secret which is reconstructed with keys of nearby neighbor nodes.

- Warn all nodes in the network when there is an intrusion attack in network.

- Send private key over a man-in-middle node by our frame to increase range of our protocol.

In order to reach these goals, we combine our protocol and scheme into a multi-thread program with the flow graph as shown in **Figure 6**.

Figures 7 and **8** show an example of development with three nodes; node 1 is within the communication range of the others, but the distance between node 0 and node 2 is too long to establish a link.

To prevent man-in-middle attack as we have mentioned before, we encrypt the node's keys each time they are transferred to the other node, so there is a problem on how the requesting node can receive exactly that key with this method. We add a

Source MAC (6bytes)	Destination MAC (6bytes)	Protocol (2bytes)	Data header (1bytes)
Hop count (1byte)	Delay time (8byte)	Data length (1byte)	Data

Figure 5.
Proposed security protocol header frame.

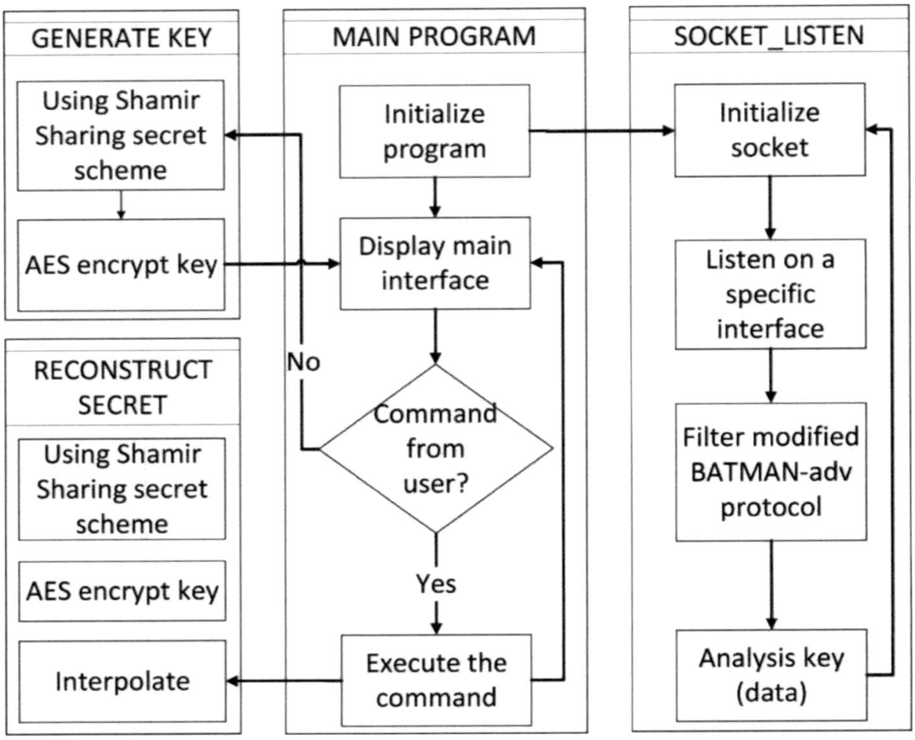

Figure 6.
Proposed program flow graph.

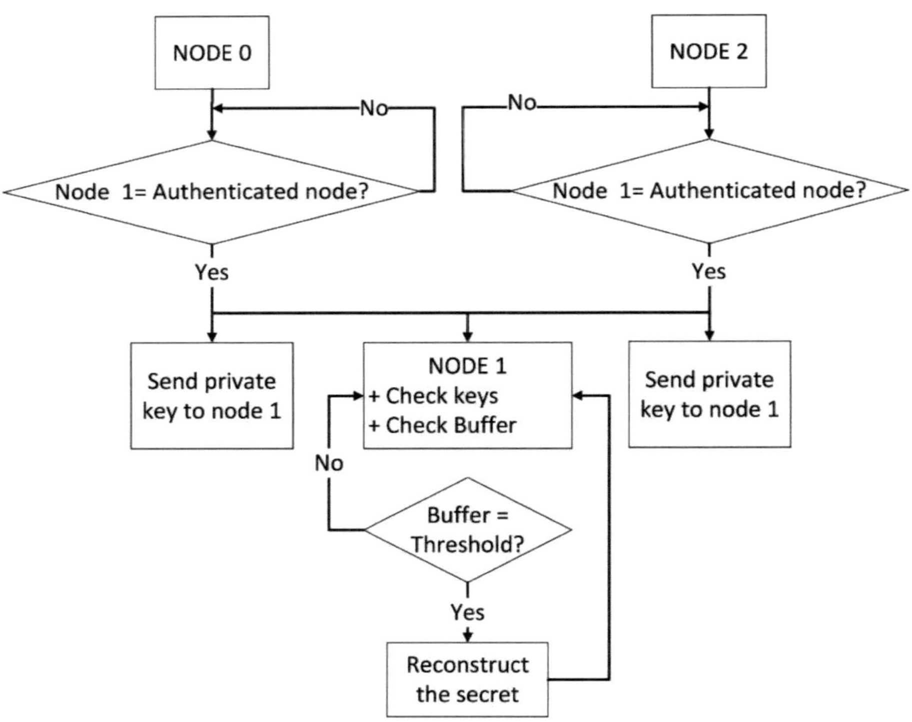

Figure 7.
Secret reconstruction of node 1.

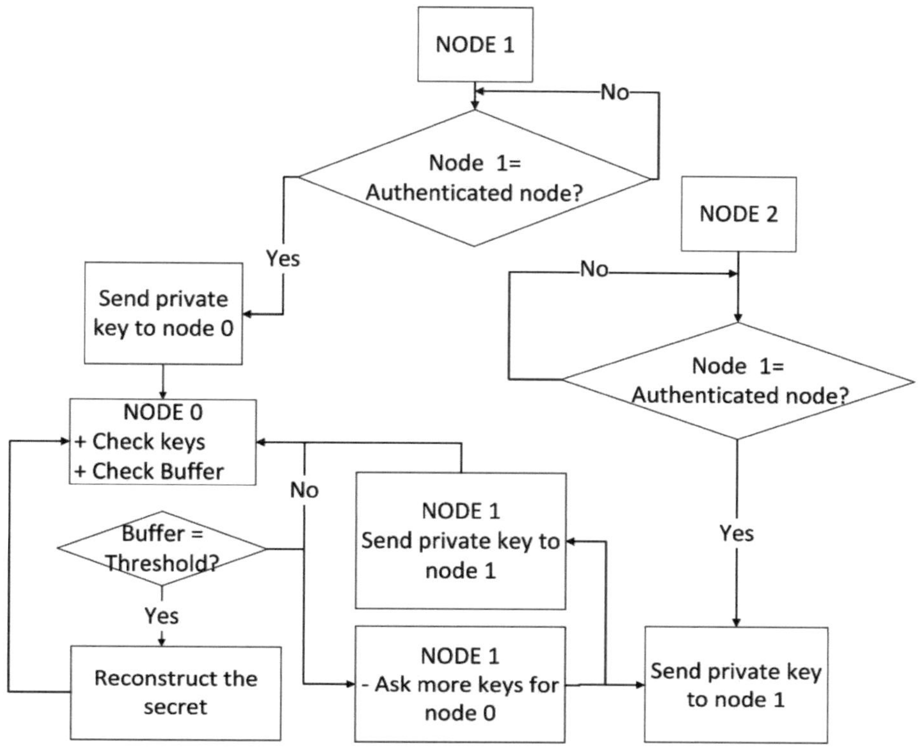

Figure 8.
Secret reconstruction of node 0.

new field "hop count" to our header frame. This field is used for counting the number of nodes which will help the packet from the requester to be transferred to the destination. Then the destination will take that value to encrypt its key <hop count> time s before sending it to the source who requested for more keys. The "key used for encrypting key" is also generated by real-time method; it must be known by all nodes between source and destination to ensure that the decryption on those nodes is correct. Another issue is the case that an intrusion node tries to decrypt more than one time to get the value of encrypted key. In order to solve this issue, we set hop count value equal to 0 so the intrusion node does not know how many times it has to decrypt for getting the key. Only the owner of the key knows this hop count value. The process of the key encryption with hop count values is shown in **Figure 9**.

Because of the delay when transferring the data, it is hard to synchronize the nodes in WMN with our real-time method; the encryption will be incorrect with only NTP system. Thus, we use a buffer at each node in WMN and a field to tell the delay time the request packet sends from the source to destination. When the destination receives the frame, it can send the key it generated at the time the source sends out its request packet; the buffer has responsibility to record all the key value in a minute latest. Therefore, if the delay value in WMN is over 1 minute, requester cannot receive the key from destination. **Figure 10** shows how the buffer and delay time field work.

5.3 System implementation

We consider a WMN consisting of sensor nodes that can work as base stations which will help non-mesh clients communicate together (**Figure 11**). In this experiment,

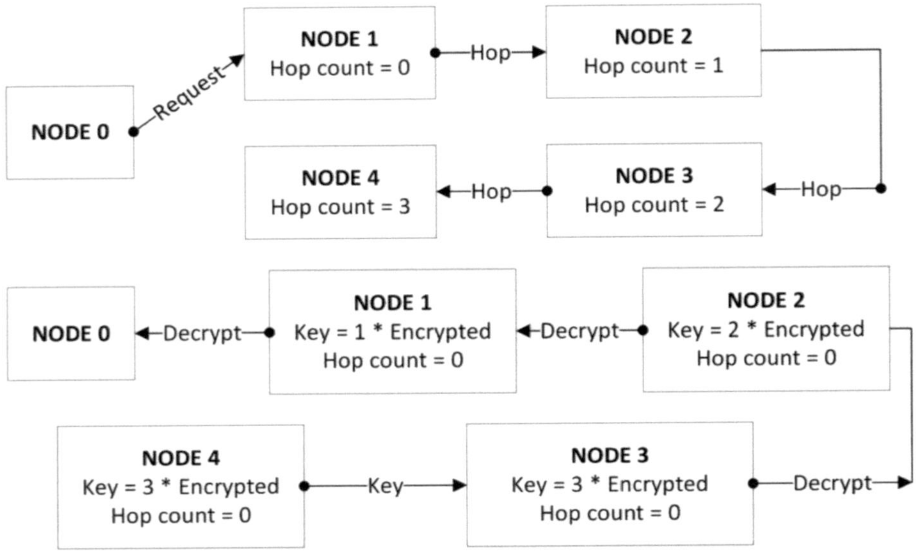

Figure 9.
Encryption of the key with hop count value.

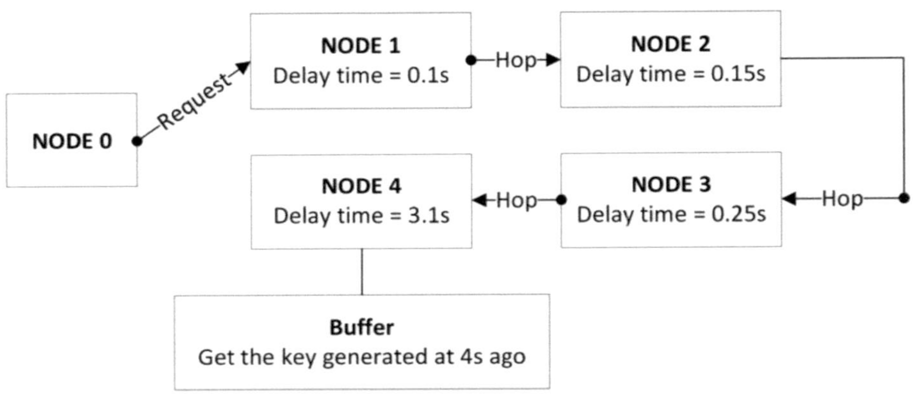

Figure 10.
How the buffer and delay time field work.

we bridge wireless local area network interface and mesh network interface together instead of putting WLAN behind the firewall in order to make our work easier, because we only check if our scheme works perfectly on data link layer not on network layer.

Our test security program has already been installed on every sensor node, and we put them in distance. As our scheme, node 1 receives private key from nodes 2 and 3, combined with its key to reconstruct the original secret which is used to encrypt data. This encrypted data is sent to one of the clients of node 4; after decrypting, we compare the decrypted with the original data to see if our scheme works completely. Another parameter which we have to check is secret after reconstruction. We will list all those parameters in the next section.

5.4 Results

In this experiment, we set secret—public key of WMN equal exactly to minute value of UTC time zone. Therefore, the time on every sensor node must be set at the

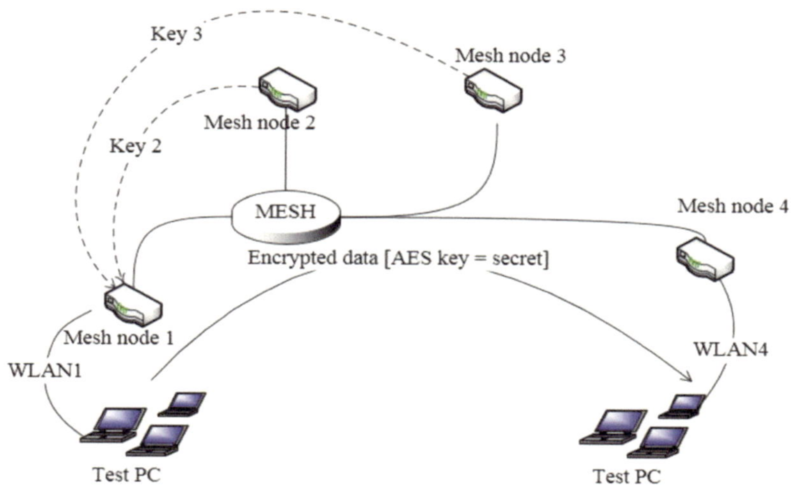

Figure 11.
System model.

Time	Key 1 (Hex)		Key 2 (Hex)		Key 3 (Hex)		Secret(Dec)
19:45	96	C4	85	F4	26	18	64
19:51	EC	C4	FB	F4	E5	18	70
22:06	8C	C4	6C	F4	D5	18	28

Table 1.
Secret reconstruction analysis.

same value. **Table 1** shows the keys (decrypted keys) collected by node 1 needed for secret reconstruction analysis at the different time. We put a simple function for our experiment secret as follows: Secret = Hour value + Minute value. We executed our program on three nodes in this experiment. This program shows us the value of generated key of each node, the number of bad nodes this security detected, and the time and the secret which was reconstructed at that time. **Figure 12** shows all those results of all three nodes at the time 22:06. Each node has a different private key from the others based on its MAC address. But all of them have the same secret at a certain time.

Secret is reconstructed exactly as the origin with at least three private keys of WMN. Therefore, we can run to the next step—data encryption—and original and decrypted data are the same in both transmitter and receiver if it works correctly, in our case are clients of nodes 1 and 3.

After reconstructing the secret completely, data which are sent from node 1 will be encrypted by this secret. Both encrypted data and decrypted data are shown. This data is captured at one of the non-mesh clients of wireless local area network node 4. We had also checked if the keys are secured when they are transmitted in our model (four nodes with maximum hops equal to 2). Then, we used an external node which worked as a monitor node to capture the raw package to check if the keys are encrypted.

We tested our methods to face types of attacks mainly in general wireless network and in particular wireless mesh network.

Firstly, we test our network model with eavesdropping attacks. We use ESP8266 kits to collect all the data from our network; the entire data was encrypted; we also tried to collect private keys from authenticated nodes in this network to reconstruct

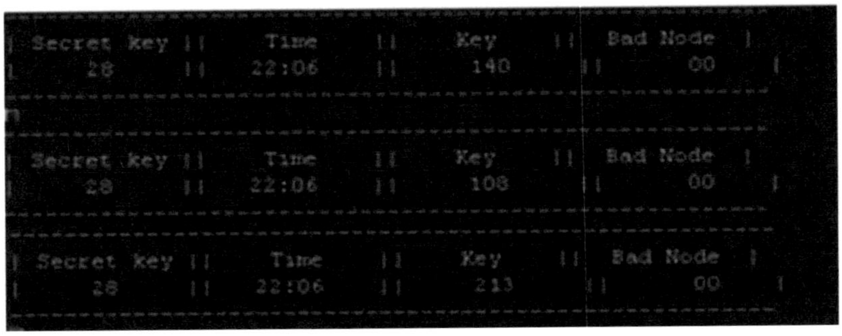

Figure 12.
Experimental figure.

the secret key, but all keys which are transferred in this network had been encrypted with the hop count parameter we had discussed before. To resolve this issue, an attacker needs to decrypt those keys with the pairs of MAC address and hop count parameter, respectively, in this kind of network. Even if attackers can decrypt all keys, they will face with the problem that the keys of our network model are constantly changed over time.

Secondly, we tested our model with many kinds of active attacks, because the nature of the connection on the layer 2 of the original BATMAN protocol has already been pretty tight so almost the active attacks up to this model are neutralized, so the impacts of them are only small impacts on single node and easily detected by our protocol when there are abnormal signs from any nodes in our model.

Next, we tried to use jamming attack to our model. Unfortunately, we have not handled this kind of attack. Therefore, in the near future, we will develop our model to overcome this drawback.

Finally, let see how our model handle the man-in-middle attacks. Because we use real time mainly in our protocol for generating keys, reconstructing the secret, and also detecting abnormal nodes. Therefore, any man-in-middle attacks without being synchronized in real time or do not have the ability to interact with the other authenticated nodes in the specified period that we mentioned in the previous section are defined as abnormal node.

To sum up, attackers only can strike this network model if they know how the protocol works. However, it requires a process to collect, decrypt, synchronize, and analyze accurately complex data from the attacked nodes.

Table 2 shows a comparison between our algorithm and the others over the security reliability criteria. Our proposed algorithm can prevent many types of attacks which we have discussed before—some of them cannot be prevented by the other algorithms.

The original Shamir's algorithm [5] has the weakest security reliability in this table because it only prevents attacks focusing on eavesdropping data. Similarly, SSKM [7] and HSTMK [8] are capable of defending eavesdropping data attacks, but they use two different methods to keep the key materials confidential. SSKM is an improvement of SSSS [5] by using a discrete logarithm algorithm to exchange the keys in a secure channel, while HSTMK takes advantages of separated sub-phases to anticipate the deletion of master secrets. The more resources are consumed by the network models, the more the number of nodes is increased. Consequently, the scale of the model deployed by this method is limited. Therefore, in order to avoid this problem, our algorithm mainly focuses on extending the scale of the network model with the custom protocol using minimal buffer on each node that we mentioned in sections above.

	Proposed algorithm	SSSS [5]	SSKM [7]	HSTMK [8]
Prevent attacks	- Eavesdropping keys and data - De-authentication attack - DoS attack - Replay attack - Man-in-middle attacks	- Eavesdropping data	- Eavesdropping keys and data	- Eavesdropping keys and data

Table 2.
Security reliability comparison.

6. Conclusion

This chapter presented four key management schemes and also security protocol for WMNs. First, Shamir's scheme was a popular method which was used for distributing the keys. The second scheme called SSKM was an improvement of Shamir's scheme by generating different keys in different period as well as using discrete logarithm algorithm to transport the key in a secret way. The third method called HSTMK utilized a key negotiation routine to solve the problem of compromised node. This one also divides initialization phase into four sub-phases to reduce the time requirement and increase the security level. Finally, in our scheme, we establish secured communication sessions between nodes so they can hide their private keys from the other except the requester. That means not only data but also keys were encrypted by combining our scheme with AES encryption. We also use the real-time value to constantly change each node's private key. This has caused great difficulty for anyone who wants to find out private keys of WMN. Comparing with existing security protocols and schemes shows that our scheme is simple to deploy, and it has a better security.

There remain some problems that should be addressed for this security protocol. We need to reduce the amount of the calculations for the proposed protocol which is deployed on routers with small flash memory. Besides, the WMN structure needs to be improved in order to make the system model work efficiently. Thus, these considerations would be developed in the future work.

Author details

Vinh Truong Quang* and Hoa Le Viet
Ho Chi Minh City University of Technology (HCMUT), Vietnam National University, Ho Chi Minh (VNU-HCM), Vietnam

*Address all correspondence to: tqvinh@hcmut.edu.vn

References

[1] Akyildiz F, Wang X, Wang W. Wireless mesh networks: A survey. Elsevier Journal of Computer Networks. 2005;**47**(4):445-487

[2] Siddiqui MS, Hong CS. Security issues in wireless mesh networks. In: Proceedings of the International Conference on Multimedia and Ubiquitous Engineering (MUE'07). New York: IEEE Press; 2007. pp. 41-47

[3] Agarrwal S, Gupta N. Authentication and key management in wireless mesh network. MIT International Journal of Computer Science & Information Technology. Aug. 2012;**2**(2):70-74

[4] Gao L, Chang E, Parvin S, Han S, Dillon T. A secure key management model for wireless mesh networks. IEEE AINA. 2010:655-660

[5] Shamir A. How to share a secret. Communications of the ACM. 1979;**22**(11):612-613

[6] Li F, Xin X, Hu Y. Key management in ad hoc networks using self-certified public key system. International Journal of Mobile Communications. 2007;**5**(1):94-106

[7] Lan Y, Wu C, Zhang Y. A secret-sharing based key management in wireless sensor network. 4th IEEE International Conference on Software Engineering and Service Science (ICSESS). 2013

[8] Gandino F, Ferrero R, Montrucchio B, Rebaudengo M. Fast hierarchical key management scheme with transitory master key for wireless sensor networks. IEEE Internet of Things Journal. 2016

[9] Singh A, Awasthi AK, Singh K. Lightweight multilevel key management scheme for large scale wireless sensor network. In: International Conference on Computing for Sustainable Global Development (INDIACom). 2016

[10] Zhu S, Setia S, Jajodia S. Leap+: Efficient security mechanisms for large-scale distributed sensor networks. ACM Transactions on Sensor Networks. Nov. 2006;**2**(4):500-528

[11] Hoa LV, Vinh TQ. Real-time key Management for Wireless Mesh Network. Journal of Telecommunication, Electronic and Computer Engineering. 2018;**10**(2-6):13-18

[12] Whittaker ET, Robinson G. Lagrange's formula of interpolation. In: The Calculus of Observations: A Treatise on Numerical Mathematics. 4th ed. New York: Dover; 1976. pp. 28-30

Chapter 5

Digest: A Biometric Authentication Protocol in Wireless Sensor Network

Faezeh Sadat Babamir and Murvet Kirci

Abstract

Since the security of biometric information may be threatened by network attacks, presenting individual's information without a suitable protection is not suitable for authorization. In traditional cryptographic systems, security was done using individual's password(s) or driving some other data from primary information as secret key(s). However, encryption and decryption algorithms are slow and contain time-consuming operations for transferring data in network. Thus, it is better that we have no need to decrypt an encrypted trait of an enrolled person, and the system can encrypt the user trait with the user's passwords and then compare the results with the enrolled persons' encrypted data stored in database. In this chapter, by considering wireless sensor networks and authenticating server, we introduce a new concept called "digest" and deal with its efficiency in dealing with the security problem. A "digest" can be derived from any kind of information trait through which nobody can capture any information of primary biometric traits. We show that this concept leads to the increase of the accuracy and accessibility of a biometric system.

Keywords: biometric authentication, wireless sensor networks, system security, cryptography, digest

1. Introduction

A biometry of a person is a physical/logical property that is obtained from trait of the person. Since traits, details vary from person to person and no two people have the same trait, so the trait can be used as an ID in authentication/identification systems. Trait is a biological property of a person like fingerprint, retinal, iris, deoxyribonucleic acid (DNA), etc. The biometric authentication system collects traits of legitimate persons and stores them in database safely in order to use for identification of them in verification time, access time to data or place, etc. Moreover, implementing an authentication system through biometric data can create a secure guarded port for secure data or a place. This biometric security system is a lock and capture mechanism for access control [1]. In order to develop such a system, traits of legitimate persons should be scanned and stored in a safe database. When biometric security system is activated for authentication, it verifies and matches traits of persons with ones in database [1].

Wireless sensor networks (WSNs) are general networks that are employed in several applications, including military, medical. In all these cases, data security and energy usage are the determining factors in the performance of critical applications. Consequently, methods of protecting and transferring data to the base station are very important because the sensor nodes run on battery power and the energy available for sensors is limited [2].

In order to implement a flexible biometric security system, we need a favorite channel for transmitting information/data. This channel should be a safe and quick passage to transmit biological traits information/data. Since most of the time, accessing secure channel is costly or impossible, we would use a WSN channel for connecting capturing equipment such as scanner to DB. Obviously, this kind of network is not an enough safe passage for transferring highly secure information/data, because an enemy may capture secure data being transmitted. Therefore, we should code or encrypt them such that it may be incomprehensible for others and enemies are not able to abuse them. This process could be done by integrating with a biometric cryptography algorithms and WSNs [3].

Moreover, we use cryptographic algorithms for raw highly secure information to convert them to ciphertext. This task provides security as well as privacy.

Current authentication systems mostly are based on ID and password authentication system. Password is a combination of characters, numbers and letters that should be renewed in certain periods to prevent unauthorized people accesses. In order to provide an almost perfect secure system, a biometric security system can be implemented for authentication. But as mentioned above, the main problem is sending and receiving secure data/matching result through unsafe network. It means that network security should be considered as part of security performance for evaluation of security level of a biometric security system [3].

In this paper, we investigate a biometric security system proposed in [4] in WSNs. It saves a print of individual biometric traits through especial framework called "digest," which is output of a one-way function. This framework supplies perfect security without carrying out any encryption or decryption processes. Therefore, it would be a good selection for privacy preserving of users who wish to be authorized through a WSN. In order to make highly memory performance homomorphic property is utilized. This issue improves the algorithm energy consumption in WSN. Finally, Hamming distance measurement is used to compare stored data with newly created data to make decision of matched or mismatched in based node.

2. Related work

There are many studies that present power complexity efficiency methods in wireless sensor networks. These studies applied natural algorithms including genetic algorithm to find best method for transferring data [3–5].

The primary authentication mechanism is fingerprint whereas it is currently being pushed by the majority of smartphone/personal computer vendors. This solution is so simple due to the fact that our fingerprints could be obtained from everywhere that we were and touched before [6, 7]. Therefore, utilizing some individuals features are recommended to be used as a standalone authentication approach. Most of the smartphone vendors install an additional camera to obtain the fingerprint [8].

Key binding algorithm is used in [9, 10] for fingerprint matching system. Moreover, a cryptographic key will be bind with the user's fingerprint images at the time of enrolment.

Davida et al. [11, 12] proposed the iris based biometric for, authentication process. Moreover, binary representation of iris texture, called IrisCode [13] is considered. Also the Hamming distance compared the input and database template representations with a threshold to determine the matching result.

Monrose et al. [14] combined passwords with keystroke biometrics in secure way. Their technique was inspired by password "salting." Disadvantage of this method is that it only adds about 15 bits of entropy to the passwords. This leads marginally security. In [15, 16] they made some minor modifications to their primary work. They applied voice biometrics instead of keystroke. Tuyls et al. in [17, 18] supposed that all template are noise-free of a biometric identifier. Thus, they used them directly in to generate a secret named helper data W.

Juels and Wattenberg Davida et al.'s methods [11, 12] to tolerate variance in "fuzzy commitment" scheme [19]. This provides more strong security. Juels and Sudan [20] showed the security of the fuzzy vault scheme in an information-theoretic sense. Clancy et al. [21] extended Juels and Sudan [20] work. Moreover, they used "fingerprint vault" for multiple (typically five) fingerprints of users.

Michelin et al. [22] proposed the use of the smartphone's camera for facial and iris recognition by the decision-making using the cloud. Another work on biometric authentication for an Android device [23] showed an increased level of higher task efficiency achieved using various solution. In [24], authors studied the usability and practicality of biometric authentication in the workplace and concluded that the ease of technology utilization and its environmental context play a vital role while the integration and the adoption will always incur additional and unexpected resource costs.

The gesture-related user experience research conducted in [25–28] showed that security and user experience do not necessarily need to contradict each other. This work also promoted pleasure as the best way for fast technology adoption. In [26], authors addressed the usability of the ECG solution for authentication and concluded that the application of ECG is not yet suitable for dynamic real-life scenarios.

3. The protocol

Here, we explain the proposed biometric cryptosystem [4] based on Finite Composite order group as well as a figure that clears logical relationship between important parts of the system (**Figure 1**). The security degree of the system depends on a hard DLP [29].

For a high security level (with selection of very large factors), factoring N (if N = nm such that (n, m) are coprime numbers) is impossible. Disadvantage of this technique is that performing group operation for large composite groups is slow leading to complicated operations. This system is based on a special *generator* to resist many attacks making the system faster. Below, we explain steps of the proposed biometric cryptosystem.

3.1 KeyGen(π)

Let d be a security parameter of the system. Let m, p and q denote very large random prime numbers in which $n = pq$ and $N = nm \therefore n < m \;\&\; (n, m) = 1$. We define m and n as modulus for biometric trait. Also we know that $\varphi(N) = \varphi(mn) = \varphi(m)\varphi(n) = \alpha\beta$.

Let $X \equiv a \bmod m$ and $X \equiv b \bmod n$, $(m,n) = 1$, $(a,b) \geq 0$, $(\forall X < \alpha)$. According to the Chinese theorem [10, 11], we have:

$$X \equiv N_1 s_1 r_1 + N_2 s_2 r_2 \;(mod\; N) = m\,m_n^{-1} b + n\,n_m^{-1} a \bmod N \tag{1}$$

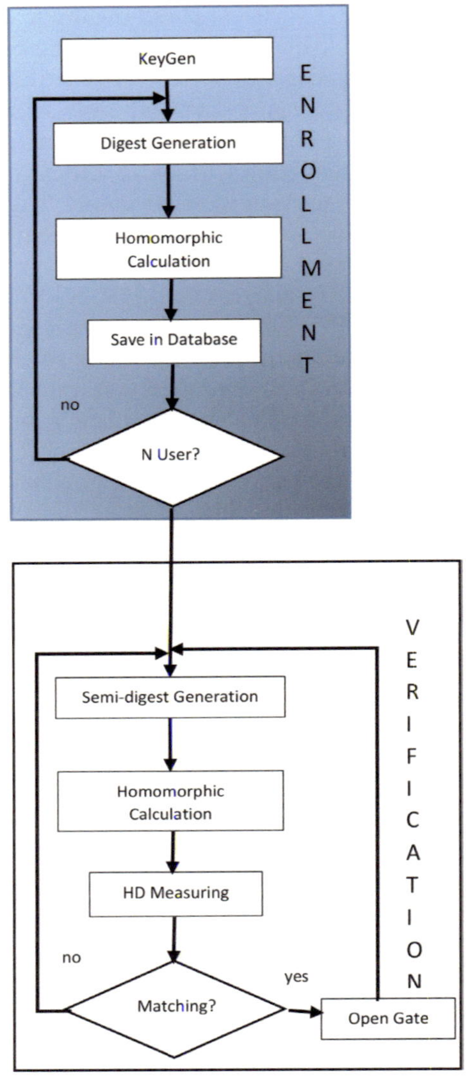

Figure 1.
The biometric authentication cryptosystem.

In Eq. (1) m_n^{-1} is the inverse of $(m \bmod n)$ and n_m^{-1} is the inverse of $(n \bmod m)$. Also $v = m m_n^{-1}, u = n n_m^{-1} \therefore (u, \alpha) = 1$.

We choose public system parameters as $< g, m, n, N, u, \alpha >$, and master secret key as $MSK = <\varphi(N)>$. Where G be a cyclic group with generator $g \in \mathbb{Z}_N$ and $ord(g) = \alpha$. Also, β and so $\phi(N)$ should kept secret.

3.2 Enrollment (UK, SK, X)

In this phase, we measure biometric trait (X) to obtain value $R = [X - bv] \bmod N$. Then system calculates k_{client} from Eq. (2), and saves this value to the memory.

$$D(X) = k_{client} = g^{R\beta} \bmod N \therefore \beta = \varphi(n) \quad (2)$$

For every client and erases, all values except k_{client} of client, will kept in the memory of the system for verification process. If everyone access to k_{client}, she/he cannot obtain no information about X or R.

3.3 Verification (UK, X′)

In authentication time, client calculates following equation:

$$h = Xu \bmod N \rightarrow D = g^{h+r\alpha} \bmod N \therefore r \in \mathbb{Z}_N^*, \alpha = \varphi(m) \tag{3}$$

In Eq. (3), r is a random number. Client sends D to the system for verification process. System receives D and verifies:

$$\mathrm{D}_{\mathrm{client}}^{\beta} \bmod \mathrm{N} \xrightarrow{?} \mathrm{k}_{\mathrm{client}} \tag{4}$$

Correctness: we now describe that how verification performs efficiently. From Eq. (4), we have:

$$D^{\beta} = \left[g^{h+r\alpha} \bmod N\right]^{\beta} = \left(g^{h\beta} \bmod N\right)\left(g^{r\alpha\beta} \bmod N\right) \tag{5}$$

According to the Euler's totient function [10], the Eq. (5) equals to Eq. (6):

$$\rightarrow \left(g^{h\beta} \bmod N\right) \times (1) = g^{h\beta} \bmod N \xrightarrow{?} k_{client} \tag{6}$$

Homomorphic verification: the scheme turns out to be useful in homomorphic verification over an additive group, i.e., if $D(h)$ be randomized biometric digest $X \in \mathbb{Z}_N$, with respect to the public parameter N, we have Eq. (7):

$$D(h_1).D(h_2) = D[(h_1 + h_2) \bmod N] \therefore \forall h \in \mathbb{Z}_N \tag{7}$$

HD measuring (M, D, M′, D′): the protocol check HD of parameter of Eq. (8), with all one in database along with their mask vectors. Note that:

$$HD\left(M, D, M', D'\right) = \frac{\left\|\left(D \oplus D'\right).M.M'\right\|}{\left\|M.M'\right\|} \tag{8}$$

Matching (HD, ι): Now the protocol compare obtained value to make final output according to Eq. (9)

$$\text{result} = \begin{cases} \text{matched HD} \leq \tau \\ \text{mismatched o.w.} \end{cases} \tag{9}$$

The protocol includes two main phases: enrolment and verification. Every user should be enrolled through entering his/her biometric features using available instruments in the enrolment phase [30]. These instruments capture images and then process them to output vectors of *feature* and *mask* to cover errors as possible and send them to the enrollment algorithm [4, 31].

The protocol does not save original information in database. Instead, the protocol keeps the information in cache for just some seconds in order to process it using mathematical one-way functions and convert it to different data with different formats and natures. The final processed data *Digest*; Digests are values that nobody even system itself can identify the owner and the biometric property of the corresponding digest. Different digests of client will be fused with homomorphic operation. Additionally, he/she cannot misused available digests, because at authentication request time or online mode, system accept just semi-digest data as input that needs one more processing step to output digest [4].

After generating all of fused digests, all of primary information is safely erased from the cache memory and the digest is transmitted to the system database. The database is set of all original digests whose owners and biometric properties are unknown. Hereafter, if an individual wants to enter the system, the system will be able to identify him/her correctly as an authorized/unauthorized client [4].

After completing the enrollment phase and enrolling clients' digests, the system runs verification process, i.e., it enters the verification time of the protocol. An individual who request for authentication, enters his/her biometric information and the system captures it, process it to make fused semi-digest [4].

From now one, the protocol starts comparing algorithms. It firstly combines semi-digest with the secret parameter of the system to generate the corresponding digest. This digest will be compared with available digests in the database. This matching will be carried out using computation of the Hamming Distance measure of four parameters: (1) the stored digest, (2) its mask vector, (3) the new digest, and (4) its mask vector. If the obtained HD is fewer than value of threshold τ, the client's identification has been matched [4].

4. The protocol analysis in WSNs

In this section, we compare the scheme [4] with those of [9–12, 14, 17, 20, 21]. Moreover, we review properties of the protocol that were performed on a single core of an Intel®, Pentium® D, 3:2Ghz processor, using MATLAB R2016a. Also we used Miracle library in binary fields [32] for some mathematical operations.

In this case, we installed the Java Genetic Algorithm Package (JPAC) to test the algorithm in a manner consistent with prior studies. Next, we utilized OMNET++ to trace the movement of the nodes in a virtual environment.

Algorithm	Biometric representation	Classification	Privacy preservation	Practicality	Sensitivity to invariance	Security	Efficiency in WSNs
Murvet et al. [4]	Fingerprint, iris	G	H	H	L	H	L
Soutar et al. [9, 10]	Fingerprint	R	H	M	H	U	M
Davida et al. [11, 12]	Iris	G	H	H	L	U	L
Monrose et al. [14]	Keystroke, voice	G	H	H	H	M	M
Linnartz et al. [17]	No evaluation	G	H	L	L	H	L
Juels and Sudan [20]	No evaluation	G	H	H	L	H	H
Clancy et al. [21]	Fingerprint	G	H	H	M	H	H

Table 1.
Comparison between various algorithms.

In **Table 1**, a comparison between various algorithms in WSNs: the proposed scheme in [4], Soutar et al. [9, 10], Davida et al. [11, 12], Monrose et al. [14], Linnartz and Tuyls [17], Juels and Sudan [20], and Clancy et al. [21] are given. The third column in **Table 1** indicates the key release (R) or key generation (G) classification. Column "Practicality" deals with the complexity of the algorithm. Last column shows the efficiency of algorithms in WSNs.

The protocol operates based on new concept *digest* [4] that leads to reduce time complexity of the proposal compared to schemes that already used encryption and decryption process.

This concept also improved efficiency of identification operation in cost and time as well as it is safe enough to customize for any application.

5. Conclusion

Wireless sensor networks are flexible and useful networks for securing critical data through biometric authentications. However, they are powered by nodes equipped by the limited capacity batteries. On the other hand, biometric authentication brings greater convenience to users than other authentication systems. This method can perfectly protect legitimated users and data against internal malicious and external frauds. Moreover, this measures and analyzes user's unique information for automatically recognizing user's identification. The first five most common traits are fingerprint, hand, eye/Iris, face and voice that would be transmitted through WSN. In this study, we utilized Iris and fingerprint to make a strong biometric authentication system in WSN. The scheme proposed in [4] was the more efficient in terms of applicable efficiency in WSN in comparison with similar studies. As a future work, the system will be able to operate in any networks by applying property of "Boolean identification." Further, by studying other difficult problems, we will improve this study to gain linear time efficiency. These new properties help networks to transmit data securely and efficiently in any sensitive network.

Author details

Faezeh Sadat Babamir* and Murvet Kirci
Istanbul Technical University, Maslak, Turkey

*Address all correspondence to: babamir@itu.edu.tr

References

[1] Jain AK, Ross A, Pankanti S. Biometrics: A tool for information security. IEEE Transactions on Information Forensics and Security. July 2006;**1**(2):125-143

[2] Norouzi A, Babmir FS, Zaim AH. A novel efficient routing protocol in wireless sensor network. Wireless Sensor Network Journal. October 2011;**3**(10):341-350

[3] Reid P. Biometrics for Network Security. Boston, USA: Pearson Education Inc.; 2004. ISBN: 0131015494

[4] Kirci M, Babamir FS. A digest based method for efficiency improvement of security in biometrical cryptography authentication. In: IEEE International Symposium on Computer Science and Software Engineering; 2017

[5] Norouzi A, Babamir FS, Zaim AH. An interactive genetic algorithm for mobile sensor networks. Studies in Informatics and Control. 2013;**22**(2)

[6] Jain A, Bolle R, Pankanti S. Biometrics: Personal Identification in Networked Society. Vol. 479. Berlin, Germany: Springer; 2006

[7] Maltoni D, Maio D, Jain A, Prabhakar S. Handbook of Fingerprint Recognition. 2nd ed. Berlin, Germany: Springer; 2009

[8] Le C. A Survey of Biometric Security Systems. A Report. USA: Washington University in St. Louis; 2018

[9] Soutar C, Roberge D, Stojanov SA, Gilroy R, Vijaya Kumar BVK. Biometric encryption using image processing. In: Proceedings of SPIE, Optical Security and Counterfeit Deterrence Techniques II; Vol. 3314; 1998. pp. 178-188

[10] Cavoukian A, Stoianov A. Biometric encryption. In: Nichols RK, editor. ICSA Guide to Cryptography. New York: McGraw-Hill; 1999

[11] Davida GI, Frankel Y, Matt BJ. On enabling secure applications through off-line biometric identification. In: Proceedings of the 1998 IEEE Symposium on Privacy and Security; 1998. pp. 148-157

[12] Davida GI, Frankel Y, Matt BJ, Peralta R. On the relation of error correction and cryptography to an offline biometric based identification scheme. In: Proceedings of the Workshop Coding and Cryptography (WCC'99); 1999. pp. 129-138

[13] Ang R, Safavi-Naini R, McAven L. Cancellable key based fringerprint templates. In: Australasian Conference on Information Security and Privacy; 2005. pp. 242-252

[14] Monrose F, Reiter MK, Wetzel S. Password hardening based on keystroke dynamics. In: Proceedings of the 6th ACM Conference on Computer and Communications Security; 1999. pp. 73-82

[15] Monrose F, Reiter MK, Li Q, Wetzel S. Using voice to generate cryptographic keys. In: Proceedings of 2001: A Speaker Odyssey, Speaker Recognition Workshop; 2001. pp. 237-242

[16] Monrose F, Reiter MK, Li Q, Lopresti DP, Shih C. Toward speech-generated cryptographic keys on resource constrained devices. In: Proceedings of the 11th USENIX Security Symposium; 2002. pp. 283-296

[17] Linnartz J-P, Tuyls P. New shielding functions to enhance privacy and prevent misuse of biometric templates. In: Proceedings of the 4th International Conference on Audio- and Video-Based Biometric Person Authentication; 2003. pp. 393-402

[18] Verbitskiy E, Tuyls P, Denteneer D, Linnartz JP. Reliable biometric authentication with privacy protection. Presented at the SPIE Biometric Technology for Human Identification Conference, Orlando; 2003

[19] Juels A, Wattenberg M. A fuzzy commitment scheme. In: Proceedings of the Sixth ACM Conference on Computer and Communication Security; November 1999

[20] Juels A, Sudan M. A fuzzy vault scheme. In: Proceedings of the IEEE International Symposium on Information Theory; Lausanne, Switzerland; 2002. p. 408

[21] Clancy TC, Kiyavash N, Lin DJ. Secure smartcard-based fingerprint authentication. In: Proceedings of the ACM SIGMM 2003 Multimedia, Biometrics Methods and Applications Workshop; 2003. pp. 45-52

[22] Michelin RA, Zorzo AF, Campos MB, Neu CV, Orozco AM. Smartphone as a biometric service for web authentication. In: Proceedings of the 11th International Conference for Internet Technology and Secured Transactions (ICITST); Barcelona, Spain; 5-7 December 2016. pp. 405-408

[23] Conti V, Collotta M, Pau G, Vitabile S. Usability analysis of a novel biometric authentication approach for android-based Mobile devices. Journal of Telecommunications and Information Technology. 2014;**4**:34-43

[24] Maple C, Norrington P. The usability and practicality of biometric authentication in the workplace. In: Proceedings of the First International Conference on Availability, Reliability and Security; Vienna, Austria; 2006. pp. 1-7

[25] Aumi MTI, Kratz S. AirAuth: Evaluating in-air hand gestures for authentication. In: Proceedings of the 16th International Conference on Human-Computer Interaction with Mobile Devices & Services; Toronto, ON, Canada; 23-26 September 2014. New York, NY, USA: ACM; 2014. pp. 309-318

[26] Da Silva HP, Fred A, Lourenço A, Jain AK. Finger ECG signal for user authentication: Usability and performance. In: Proceedings of the 6th International Conference on Biometrics: Theory, Applications and Systems; Arlington, VA, USA; 29 September–2 October 2013. pp. 1-8

[27] Katz J, Lindell Y. Introduction to Modern Cryptography. Chapman and Hall/CRC; 2014

[28] Trapper W, Washington LC. Introdunction to Cryptography with Coding Theory. Upper Saddle River, NJ, USA: Prentice-Hall, Inc; 2005

[29] Babamir FS, Bayat F. Linearly Time Efficiency in Unattended Wireless Sensor Networks. Rijeka, Croatia: InTechOpen; 2012. pp. 213-226

[30] Daugman J. The importance of being random: Statistical principles of iris recognition. Pattern Recognition. 2003;**36**(2):279-291

[31] Babamir FS, Norouzi A. Achieving key privacy and invisibility for unattended wireless sensor networks. The Computer Journal. Oxford University Press; 2014;**57**(4):624-635

[32] Maltoni D, Maio D, Jain A, Prabhakar S. Hanbook of Fingerprint Recognition. 2nd. New York city, USA: Springer; 2009. https://www.miracl.com/

Chapter 6

User Authentication Based on Knowledge of Their Work on the Internet

Pavel B. Khorev

Abstract

This chapter analyzes existing user authentication methods for remote access to information systems and disadvantages of these methods. The method of multifactor authentication of users when they are accessing remote information systems, combining validation of knowledge on secret password and verification of conformity of the habits and preferences of Internet user's interests, is defined by registration in the system. Using the history of Web pages, the Internet browser creates a list of Web pages the user has visited in the past period of time. It is proposed to use the Bayesian classification for user's knowledge based on the analysis of information about Web pages visited by the user. For user authorization from someone else's computer, the user is invited to ask for additional questions to test knowledge of subject areas, which they selected during registration in the information system. This chapter defines the language and tools for implementation of the proposed authentication algorithm: the programming language PHP and the MySQL database management system (to create a database of registered users), Web-based open source application phpMyAdmin (to create and administer MySQL database management system), and the JavaScript programming language and HTML (for creating extensions for browsers receiving a list of the addresses of the Web pages visited by the user).

Keywords: user authentication, remote access, the document object model, classification of text documents, Bayesian method, PHP programming language, MySQL database management system

1. Introduction

Many Internet sites and portals (including educational institutions) should limit access to their content (commercial secrecy, personal data, intellectual property, and other sensitive information) for unauthorized users. For example, universities of distance education must provide reliable authentication of students in carrying out evaluation tasks. Financial institutions (banks) should provide access to customer accounts only after credible evidence of their authenticity.

A security user login procedure largely determines the security of an information system as a whole (and in the case of distance learning systems and the reliability of the results of implementation of the students of educational tasks). Authenticating the name of the logged in user is one of the steps in the logon process.

To authenticate users of information systems, there are well-known techniques. The first group of such methods is based on checking the knowledge of the user based on some memorized secrets (e.g., secret combination password). In the authentication process, knowledge of this secret is validated. The second group of authentication methods is based on checking the user ownership of certain hardware which becomes the subject (device), such as a smart card or USB token. The device contains the base secret of the user (e.g., its private key digital signature), which does not need to remember. Third party authentication methods are based on checking whether a user has characteristics that could not be separated from it. These characteristics, for example, can be printing finger, face, or voice. Such methods are related to biometric authentications.

But all these methods are as effective as possible in local authentication, when there is no doubt about the source of information for user authentication. If remote authentication has no such confidence, because data for verification can be provided to outsider, you want to create new authentication methods, suitable for use in remote access to information systems. These methods can be used in addition to the existing methods of authentication. Additional authentication methods, for example, are to verify a user based on the knowledge test about his preferences and competencies.

Data on user knowledge can be collected using the analysis of the content the user has visited Internet resources. Such an analysis could be based on the methods of classification of text documents.

2. Analysis of existing methods

Traditional methods of authentication (verification of secret knowledge-based reusable password and verification of biometric characteristics and devices) have a common disadvantage: the ability to intercept confirming the authenticity of the user information with its subsequent playback of the infringer to perform any action against a target system on behalf of a registered user. To improve protection against unauthorized access to sensitive information, it is usually recommended to use the two or multifactor authentication.

Consider the drawbacks of traditional methods of authenticating users of information systems in the case of remote access systems.

2.1 User authentication based on validation of secret knowledge

The main advantage of validation of knowledge-based authentication of user secret reusable password is the ease of its implementation and use. At the same time, the password authentication has many drawbacks:

- many users choose passwords that are easy enough to pick up due to lack of password length, their simplicity, and repetitiveness;

- the possibility of using the violator of readily available software tools for picking passwords;

- the ability to use social engineering techniques by the infringer (obtain the password by tricking the user); and

- the ability to "steal" the password as you type with the keyboard or intercept the password when it is sent over a computer network.

2.2 User authentication using devices

User name authentication using authentication devices is based on the uniqueness and the confidentiality of the information contained in the memory of the device. As such, information, for example, the private (secret) key of the user's electronic signature could be used. In the process of authenticating, the correctness of such key is validated using the user's public key certificate issued by a trusted certificate authority and is stored in an information system in which the user registered.

Most often the following devices are used for authentication:

- tokens that require connecting to your computer using a USB port and constituting in fact microcomputer;
- smart cards also constituting a microcomputer, but additionally requiring the use of card readers;
- passive devices (e.g., iButton or Touch Memory), which can only store information to authenticate the device owner.

For active USB devices, added protection from theft applies reusable passwords (so-called PIN-codes), the knowledge of which confirms the use of authentication devices to its rightful owner. Other advantages of authentication devices are no limits in the length and complexity of storage in the device memory and the ability to detect the fact that the device is lost or stolen and lock it in this case.

Authentication procedure using active devices may include the generation and verification of one-time passwords or occasional request response calculation (model "handshake").

But the use of authentication devices also has a number of disadvantages:

- the possibility of device failure or accidental damage;
- the additional cost issued by the registered users of the devices and their readers;
- the need for a free USB-port or additional equipment to connect your device to your computer;
- the possibility to manufacture copies of analog devices or wrongdoing or creating his software emulator; and
- the need to deploy a public key infrastructure (PKI) when using the private key of the user as electronic signature stored on his secret device [1].

2.3 Biometric user authentication

The biometric authentication checks that the user is unique and inseparable from his personality characteristics shared by physical or static (patterns of papillary lines or fingerprints, hand shape, iris and the retina of the eyes, face shape, etc.) and behavioral or dynamic (timbre, handwritten signature, tempo text input with the keyboard or keyboard "handwriting," etc.).

The advantages of biometric authentication refers the validity of authentication, user friendliness (it does not need to remember long and complex passwords or permanently carry the device authentication), and the complexity of the falsification of biometric characteristics of the offender.

The disadvantages of biometric authentication are:

- the additional cost of the equipment to read the biometric characteristics;
- storage standards of biometric characteristics in plaintext, resulting in risk of violation of the privacy of the user;
- the possibility of failure to a registered user due to an accidental large deviation of his scanned characteristic from the reference value;
- the possibility of interception of biometric characteristics when it is sent over the network.

For protection against interception of biometric characteristics and its subsequent reproduction and when the violator tries to log on to the system on behalf of others, cryptographic methods and tools can be applied. However, the use of encryption when data are transferred across the network assumes the task of managing the encryption keys. The use of a digital signature to confirm the source of biometric data requires the solution of the problem of public key certificate management devices to read such data. These causes reduced the effectiveness of the use of biometric authentication for remote user access.

Biometric authentication in Russia, now, has been started to be used to authenticate clients during their remote access to their accounts [2]. In this case, users must first register with the bank on the list, which is set by the Central Bank of Russia. To authenticate the user, the following actions are then performed:

1. entering their login and password set during registration;
2. photographing their face using camera notebook or other devices (e.g., tablet, smartphone, etc.); and
3. using a microphone, the computer speaks the text received from the authentication server and displays on the screen.

This method of authentication refers to multifactor authentication. It combines checking the knowledge secret password and authentication based on static (face) and dynamic (voice) biometric characteristics. The use of this method does not impose additional requirements to the equipment of users' computers. Specific technical solutions to this project refer to the trade secrets of its developer and financial organizations. Therefore, the effectiveness of addressing the shortcomings of biometric authentication, mentioned above, is difficult to assess.

For this reason, use a similar solution for remote user authentication information systems (e.g., universities of distance education) which appears to be unfounded so far.

2.4 The use of traditional methods of authentication for remote user access

Overall lack of traditional authentication methods for remote user access is the lack of reliable evidence of the source of data for authentication. These data can be reproduced after their "sniffing." One solution to this problem might be to establish a secure connection between a client and a server using SSL/TLS. Such a decision requires the establishment of a public key infrastructure (PKI) and certificate management [1]. This places additional requirements on the information systems and their owners. Such requirements may be redundant for distance education universities and other organizations with limited budgets.

Another possible solution would be to use the USB device of the remote users to generate one-time passwords for authentication procedure. One-time password intercepts the violator as it will not give the possibility of unauthorized access to information system resources. The use of this decision will complicate the administration of information system and will require additional expenses. Therefore, such a decision is also uncomfortable for distance education universities.

Authentication based on user testing of knowledge collected during his work on the Internet is free from these deficiencies.

3. User authentication method based on knowledge

Knowledge-based authentication often is used as a second authentication factor when using a password or user password recovery in case of loss. In this authentication scheme, the user is prompted to answer at least one additional "secret" question. But this simple schema is not free from following flaws:

- the user can forget his answer to the question;
- user response can be guessed by the infringer;
- the number of supplementary questions may not be very large; and
- answers to additional questions contain only a very small part of the knowledge of the user.

So an authentication method should be developed, which involves the collection of sufficient information. The information collected should be unique for each user, registered in the information system. It is also advisable to use the developed method that does not require any additional user action.

When designing a remote authentication method based on the knowledge of the user on the Internet, you must ensure the collection, accumulation, and use of information about the habits and preferences of the user global network. Analyzing the data of interest, habits, and preferences of Internet user may apply the analysis log of visited Web pages by this user. Among the functions of Web browsers is a function of preserving the history of visited sites and portals by the user in the appropriate journal. This function does not need to include especially constant collection of data on Internet user has visited, in the visit log saved addresses and titles of visited Web pages, as well as the date and time when they were.

Getting the user browsing history of Internet resources (scanned documents) is possible through the development of special extensions (Add-ons) for Internet Explorer and other browsers [3, 4]. However, browser manufacturers can set restrictions on the use of extensions; for example, Google Chrome, which allows installing extensions only from the shop, Chrome Web Store. Enable developer mode gives you the ability to install extensions from an arbitrary location (e.g., from the selected developer folder).

Using a known document object model (DOM) [5], it is possible to present the contents of the document (e.g., a user visited the Web page) in the form of a set of objects with certain properties. Support for this model is obligatory for all Web browsers.

In the DOM, document is presented in a tree structure. It provides a unified way to navigate through the document. This tree structure is called a node tree. Access to all the nodes can be accessed through this tree.

Using the document object model in the analysis of any user visited the Web page allows you to retrieve the value of the properties, which contains the keywords, description of the document, its title, and a list of all its internal headers list captions to the pictures (if they are available in the document). Obtaining these data provides an opportunity to analyze the document and determine the:

- set of key words;
- the number of occurrences of each of these keywords (phrases) into a document; and
- the position of the occurrences of keywords in a document.

Additionally, the results of the analysis provide an opportunity to offer the user a list of keywords (phrases) that best characterizes his interests.

For automatic document classification, visited by the user during his work on the Internet (its inclusion in one or more thematic rubrics), further analysis of the content of the document is required. You can use the following methods of classifying [6]:

- **Method of support-vector machines (SVM):** this method solves the problem by constructing a nonlinear plane separating the decision. Due to the peculiarities of nature space signs, the border decision method of supporting vectors was built, which has a high degree of flexibility in solving problems of classification of various levels of complexity.
- **K-nearest neighbors method (K-NN):** the method is based on memory usage and, unlike other statistical methods, does not require prior training, designed for classification. This method provides high efficiency, but demanding to computing resources in the stage classification.
- **Bayesian method:** this method is based on the theorem stating that if the densities of the distribution of each of the classes are known, then the search algorithm can be written in an explicit analytic form. This algorithm is optimal and has minimal error probability. In practice, the distributions of classes typically are not known. They have to be assessed (restore) on training samples. As a result, Bayesian algorithm ceases to be optimal; so as to restore the sample density is possible only with some margin of error.
- **Decision tree method:** decision tree-based classifier for category is a tree whose nodes are the terms; each edge is a labeled condition, and leaves are marked. In practice, use the binary decision trees, in which the decision of moving on the ribs is done with a simple check for terms in the document.
- **Method of neural networks:** artificial neural network is a collection of interconnected neurons. Each neuron is an elemental converter input signals at output signals. Passing on a specific set of network input signals, we get a certain set of signals to the output. A text categorizer based on neural networks is a network of elements which forms input elements presented by the terms of

the document output items submitted by categories, and links between the elements that define a dependency relationship and are marked with weights.

For remote user authentication, documents are classified and analyzed to identify those subject areas that are of interest to the user. To store the collected information, the database (DB) is used. The database will then be used for the remote user authentication.

Bayesian method is used as the proposed method of authentication. This method of classification is based on the theorem stating that if the densities of the distribution of each of the classes are known, then the classification algorithm with the minimal probability of errors can be specified explicitly. In practice, the distribution density classes not known. These probabilities has to evaluate (restore) on training sample.

Bayesian method is used when solving different tasks of information security: when spam is detected in an e-mail message, when evaluating the security of information systems, and others.

In our case, the classified document is rich in properties whose order is not important. The submission of the document was obtained by its previous analysis.

The advantages of Bayesian classification method include:

- this method allows the relatively quick classification of Web pages that must be specified when the user is authenticated;

- this method is characterized by the ease of programming;

A database containing information on behalf of the user, whose authenticity is confirmed, includes the following tables [7]:

- In the table "users," set attributes (columns) as id of the user, his name ("login"), the hash value of the password, the e-mail address of the user, sign mandatory password change at next logon, and date and time of the last logon user. It contains information about the users registered in the information system.

- In the table "interests," set the id attributes of interest and its name. It contains information on those subject areas that represent the interests of the user.

- In the table "users_interests," set columns as the id of connection user and interest, user id, id of interest. Information from this table links a user and his interests, identified by the analysis log of visited Web pages of the user's Internet browser.

- In the table "keywords," set the columns id keyword, keyword. Keywords are stored here, and they let you associate a document with a specific subject area.

- In the table "questions," set the columns id question, id of interest, which includes the question, the text of the question, and the user's response. The questions stored here will be asked to the user for authentication when it is not possible to analyze the history of the Web pages they have visited.

Relational database model consisting of the specified tables is presented in **Figure 1**.

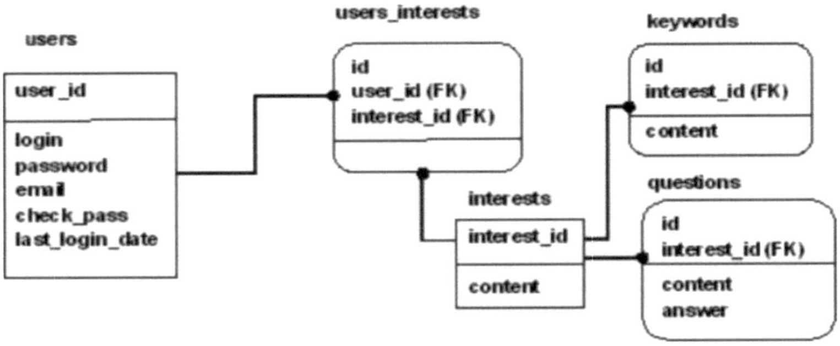

Figure 1.
The relational model DB.

When registering, the user specifies his username and password. Hidden to the user happens an analysis of visited Web pages using browser extensions. Further defines the user's interests. All data received are stored in the database. List of interests will be stored in the table "interests," associated with the "users" table—list of users of the system. Each user has an individual set of interests, so the "users" table one column will be "interest_id," which will store a list of interests of each user. The table "interests" will need at least three columns: "id" (number of entries in the table), "interest_id" (the number of the record interest in the table), and "content" (the name of interest).

In order to verify the conformity of the contents of your browser's browsing history, interests of user authentication need to be somehow mapped. Each html page can have a set of keywords, description, and header (title). After receiving a list of interests on the basis of the last visited URLs, the received data must be compared with user data stored in the database of the interests that have been entered into it after registration.

4. User authentication algorithm based on checking his knowledge

If the user tries log in into the information system from his device, it can authenticate using the following algorithm. In this algorithm, the user's browser history and his interests are identified (specified) and compared with the data received when you register a user in the system (see **Figure 2**):

1. Get a list of URLS of the Web pages contained in the log visits.

2. Retrieve tags for each Web pages from the list (its title, list of keywords, description).

3. Analysis of the information obtained to determine the user's interests.

4. Retrieving information about the interests of the user from the database that was created when its registration in the system.

5. Comparison of two sets of interests for checking user knowledge.

This method retrieves the last 1000 entries from the user's browsing history over the past 30 days. if the number of records for this period is smaller than 1000 analyses of all the log entries for the specified period of time.

Each time a remote user is authorized, the login and password are checked, as well as data analysis on its work on the Internet. Unbeknown to the user, the list of

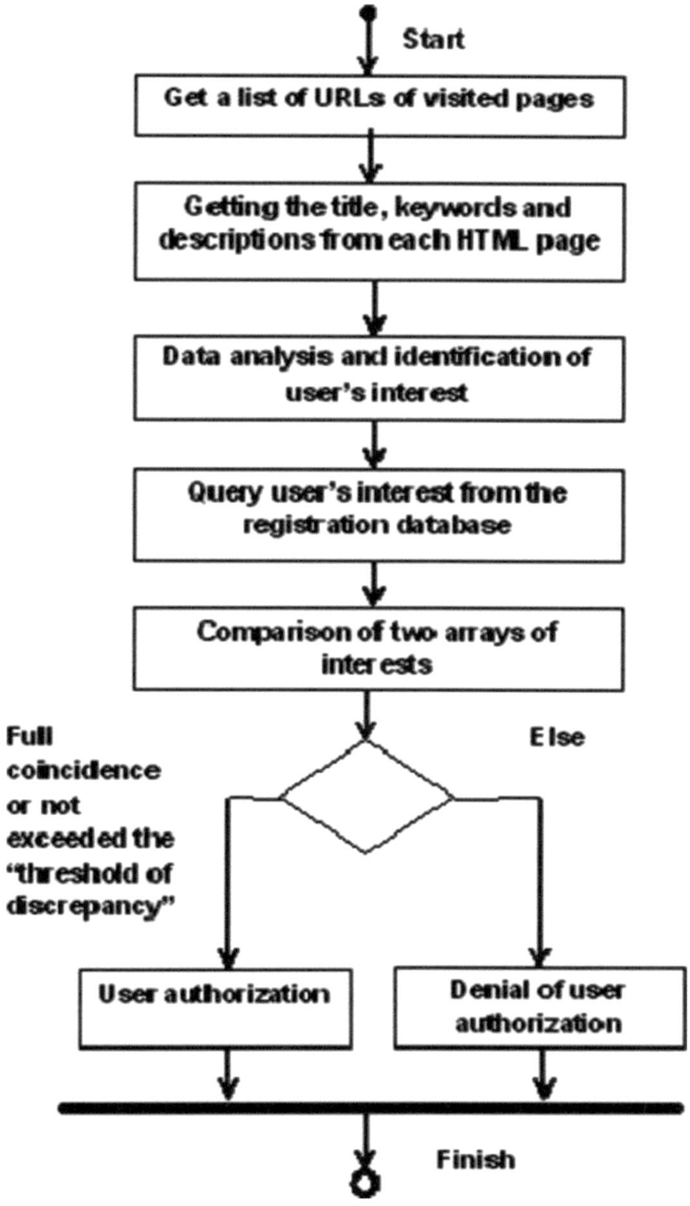

Figure 2.
The algorithm for checking the conformity of the contents of the history of the user's browser to its interests.

recent addresses of visited Web pages using Internet Explorer extensions is accessed. Further defines the user interests based on the information about the documents in history. Then the interests of the user who is authenticated are compared with those that are stored in the database. Authorization will be considered successful based on the two conditions:

- if "login" and the hash value of the password match; and
- if the difference between the interests of the user defined when authentication and retrieved from the database does not exceed the so-called "threshold of discrepancy" [8].

Let us say that from the moment of registration of the user prior to its authorization in the system, the user has actively worked on the Internet and visited the new Web pages that are not reflected in the browser's history of the Internet when registering. Then, the set of the interests of the user defined with its authorization may not match the set, which is stored in the database. Hence the use of "threshold of discrepancy" sets the maximum allowable difference between the two sets of interests. This threshold can be set by the administrator of the information system, where registering the user.

Let us say that when authorizing a user, the "inconsistency threshold" is not exceeded, and the set of certain user interests is less than his set of interests from the database. In this case, during the authorization process, the user will have to answer the questions of those subject areas that are not in the set, a specific authorization. The user is given a limited time to each response. One interest from DB corresponds to one question, and the user's incorrect answers are fixed.

Next, the user will be asked new questions of those substantive areas, the questions of which he gave incorrect answers. The maximum number of issues relevant to each interest is too limited. If you then remain relevant to the interests of user domains, the questions of which he was unable to give the correct answers, then the user authorization will not be available.

If in a set of interests that are stored in the database for each registered user, no interests, which were identified as a result of his successful authorization, these new interests are added to the database.

Application of the developed method of authentication will increase the validity of this procedure when providing remote user access to information system resources. This will reduce the potential damage from thefts of valuable information. For universities of distance education, possible loss may be associated with damage to the business reputation of the extradition documents on education for student evaluations that were falsified.

5. Methods and means of implementation

To create DB registered users, apply the programming language PHP and the database management system (DBMS) MySQL as well as Web-based open source application phpMyAdmin, designed to create and administer MySQL DBMS. phpMyAdmin allows you to administer a MySQL server, which can execute SQL-queries and view the contents of database tables.

Using phpMyAdmin, create a new database and add 5 new tables: "users," "interests," "users_interests," "keywords," and "questions."

Web browser extensions (such as Google Chrome and Mozilla Firefox) can be created using programming language (such as JavaScript) and hypertext markup language (HTML). This expansion will be used when authenticating for getting address list of Web pages viewed by the user.

Let us look at how to create extensions, for example, for Google Chrome browser. The file was originally created with the obligatory ".JSON" manifest, which contains information about the extension: extension name, version, description, version, and the location of the manifest icon in the browser address bar approx.

Example manifest file:

```
{
 "name": "Typed URL History", //the name of the extension
 "version": "1.2", //version of the extension
```

```
  "description": "Reads your history, and shows the top thousand pages you go to by
typing the URL.", //description
  "permissions": [
        "history",
        "tabs"
        ],
  "browser_action": {// the extension will have an icon next to your address bar
  "default_popup": "typedUrls.html", //the title of the html page that will be
              //displayed when clicking on the icon extension
  "default_icon": "url.png" //the name of the image that will be used as the icon
         },
  "manifest_version": 2 //manifest version
}
```

After you create a manifest, they are created with HTML and JS-files: "typedUrls.html" (HTML page that describes the type of window that is displayed after clicking on the icon extension) and "typedUrls.js" (the file that implements the collection of information about the user's browser log). For the implementation of the algorithm, the following functions were created:

- function showURLs(historyItems) (gathers a list of URLs from the user's browser history); and
- function showHistory() (displays the collected history pages for a specified period of time).

To invoke the necessary functions, event handler "addEventListener" is used.

Creating such extensions when using the proposed method of user authentication allows you to automate the process of analyzing your browser history at the time of registration and authorization of the user. Users will not be required to enter any additional information for its authentication (it introduces only the "login" and password). The extension generates a list of Web page addresses. This list is passed on to the authorization service. Then this list is parsed to determine the set of user's interests (using Bayesian method) and decision on user authentication or deny his access to the system.

When implementing user authentication algorithm, two Web pages are created:

- a page with a form for data input by the user's authorization; and
- a page with a form for user registration.

The master page is considered to be an authorization form. Here you can log in if already registered, or register by clicking on a hyperlink.

On the logon page, the user is allowed to go through the procedure of authorization. If the user has not yet logged in, you can go to the registration page. On this page, the user specifies the user name ("login"), as well as an e-mail address, which will be sent with a random initial password, that will be created by the service registration. If the user has entered valid data that satisfy the conditions (the login name should be between 5 and 15 characters, containing only letters of Latin alphabet, digits, and the characters ' _ ' and '-', and e-mail address must be valid and cannot be used twice), then the user will be registered.

When a new user is authorized for the first time, it will need to change the initial password. Without changing the initial password, the user is not authorized and will be accessible only to change password page.

If a user has forgotten his or her password, he or she may recover it by using the function "forgot password?"

To exclude threats of kidnapping registered users passwords directly from a database on a Web server, passwords should be stored in a database in a hashed form. To do this in PHP, there are special functions, e.g., md5 (MD5 hashing algorithm that produces a hash value with a length of 128 bits) [9]. This function returns the result string with hexadecimal hash value.

It is possible to crack a user's password by using a special dictionary. To protect you from this attack, the password is hashed together with a random number (salt). This salt can be calculated using the PHP function uniqid. This function uses the system timer and pseudorandom number generator for maximum uniqueness and unpredictability of salt.

Impurity is added to the password when it uses concatenation operation (.) and stored in an additional field "uniqid" database table of registered users.

When registering the user list of URLs stored in an array, the Next array analysis and Bayesian classification occur in subject areas that you are interested. After their definitions, user interests are recorded in the database as follows:

1. Access the table "interest" to obtain a unique number (the "interests_id") of each user's interests.

2. Next all the unique rooms of interest are assigned to a user in table "users_interests."

3. In the field "user_id," a unique number of the user; and in the field "interest_id," a unique room of interest are recorded.

4. Thus, in table "users_interests," rows as many as the user's interests will be stored exactly, as it was determined.

When you try to log in, user input verification occurs with those that are stored in the database (the password is hashed first, and then compared with the one stored in the database password). Username and password must match exactly with those that are stored in the database.

When authorizing, a user browser extension should get a list of thousands of URLs, which he attended in the last 30 days. If their number is less than 1000, the extension will keep all the available URLs for these 30 days. Next to each URL is determined by its area of expertise (there may be several).

After categorization, the entire list of URLs of interest of the user is compared with its interests in the DB. If the difference exceeds the threshold of discrepancy, the authorization will be refused.

If the user is authenticated from someone else's computer, to authenticate it gets a list of interests. In this list, user must select the interests of the subject areas that have been identified during registration. Then the user specifies additional questions—one for each subject area (as described above).

If the remote user session duration exceeds the maximum possible period, to continue the work he would have to pass reauthorization. After a specified period of time to a user, that is, on any Web pages, page opens instead of "authorization." Such a modification is introduced in order to enhance the security of user in the system, because while you are out of the workplace, an attacker could gain access to confidential data, posing as the owner of the account records.

6. Conclusions

The principles of authentication of users based on their knowledge of their work on the Internet are identified, as well as analyzed by means of collecting such knowledge. The methods to gather and compile information about users of the Internet are analyzed, including browser history log and the DOM of an html page. The methods for solving classification tasks in relation to the interests of Internet users are also analyzed. Their advantages and disadvantages are revealed. In order to accomplish the above objective, Bayesian method was selected.

Also, authentication algorithms are developed and implemented for:

- checking the conformity of the contents of the user's browser history and its previously defined interests; and

- calculating the level of "inconsistency" and the decision to authorize a user.

Extensions for browsers such as Google Chrome and Mozilla Firefox, allowing receiving log information browser visits within a specified time period are developed as well.

Thus, the work examines the shortcomings of existing methods of authentication when accessing remote information system. The method of multi-factor user authentication does not require the user to commit additional action during authorization. This method increases the reliability of the user's authorization results compared to the password authentication.

Compared to the use of the device-based authentication method, this method does not require extra costs and does not complicate the administration of information systems due to the need for programming and accounting for issuance of authentication devices.

The application of the method described does not require creating a cryptographically secured connection between a remote user and server information system. Setting a connection involves the creation of a public key infrastructure that also complicates the administration of information system.

Application of the developed method of authentication increases the security of your information systems without the need to increase the cost of its administration. This is especially important for organizations with limited budgets, which include distance education universities.

Acknowledgements

The author expresses sincere gratitude for student E.V. Mazaeva, for making software implementation of the proposed method.

Author details

Pavel B. Khorev
National Research University "Moscow Power Engineering Institute", Moscow, Russia

*Address all correspondence to: pbkh@yandex.ru

References

[1] Smith RE. Authentication: From Passwords to Public Keys. Boston: Addison-Wesley Publishing Company; 2002. 576p. ISBN-13: 978-0201615999. ISBN-10: 0201615991

[2] Remote Identification [Internet]. 2018. Available from: https://www.cbr.ru/fintech/remote_authentication/ [Accessed: 29 March 2019]

[3] How to Add Add-ons in Internet Explorer [Internet]. 2018. Available from: https://www.wikihow.com/Add-Addons-in-Internet-Explorer [Accessed: 29 March 2019]

[4] Create Your Own Browser Extensions, Part 1. Extend Your Reach Into Chrome [Internet]. 2013. Available from: https://www.ibm.com/developerworks/library/os-extendchrome/ [Accessed: 29 March 2019]

[5] Document Object Model (DOM) [Internet]. 2009. Available from: http://www.w3.org/DOM/ [Accessed: 29 March 2019]

[6] Kuznetsov RF. Web Page Classifier Based on SVM-Multiclass [Internet]. 2006. Available from: http://romip.ru/romip2006/10_kuznecov.pdf [Accessed: 29 March 2019]

[7] Khorev PB. Authenticate users with their work on the Internet. In: 2018 IV International Conference on Information Technologies in Engineering Education (Inforino); Moscow, Russia; 2018. pp. 1-4

[8] Mazaeva EV. Method of two-factor authentication of users based on knowledge of their work in the Internet. In: Science and Education: Materials of the XII International Research and Practice Conference; Munich; November 2–3, 2016. Munich, Germany: Vela Verlag Waldkraiburg; 2016. pp. 67-69

[9] The MD5 Message-Digest Algorithm [Internet]. 1992. Available from: https://www.ietf.org/rfc/rfc1321.txt [Accessed: 29 March 2019]

Chapter 7

MAC Aspects of Millimeter-Wave Cellular Networks

Hossein S. Ghadikolaei

Abstract

The current demands for extremely high data rate wireless services and the spectrum scarcity at the sub-6 GHz bands are forcefully motivating the use of the millimeter-wave (mmWave) frequencies. MmWave communications are characterized by severe attenuation, sparse-scattering environment, large bandwidth, high penetration loss, beamforming with massive antenna arrays, and possible noise-limited operation. These characteristics imply a major difference with respect to legacy communication technologies, primarily designed for the sub-6 GHz bands, and are posing major design challenges on medium access control (MAC) layer. This book chapter discusses key MAC layer issues at the initial access and mobility management (e.g., synchronization, random access, and handover) as well as resource allocation (interference management, scheduling, and association). The chapter provides an integrated view on MAC layer issues for cellular networks and reviews the main challenges and trade-offs and the state-of-the-art proposals to address them.

Keywords: millimeter-wave systems, initial access, resource allocation, mobility management, MAC layer design, 5G

1. Introduction

Increased demands for higher data rates in wireless communication systems, along with new applications such as massive wireless access have motivated enhancing spectral efficiency by using advanced technologies such as full-duplex communications, cognitive and cooperative networking, interference cancelation, and massive multiple-input multiple-output (MIMO). As these enhancements are reaching the fundamental capacity limits, determined by the available spectrum at the sub-6-GHz bands, the millimeter-wave (mmWave) band is becoming an alternative and promising option to support extremely high data rate wireless access [1–3]. The main reason is simple: the aggregated bandwidth of most current (main) commercial wireless systems is <2% of the bandwidth available at the mmWave. Such bandwidth can easily support a Gbps data rate in most use cases.

At the moment, the main use cases of the mmWave spectrum are satellite communications, military applications, point-to-point systems, local multipoint distribution service, and short-range networks [4]. Severe attenuation of the signal at the mmWave band, especially at certain frequency bands such as 60 and 180 GHz (see Figure 1.1 in [5]), had led to a common belief that mmWave communications are suitable either for special applications with special hardware (as mentioned above) or for "whisper radios" for wireless personal area networks (WPANs) with coverage distances of a few meters [2]. However, recent studies on mmWave mobile networks have convinced academia, industry, and regulatory bodies to repurpose the mmWave band for future wireless (and even mobile) networks.

MmWave communications are particularly attractive for ultra-short range/ high rate communications and gigabit wireless applications such as wireless gigabit Ethernet and uncompressed high-quality video transmission, see **Table 1**.

Usage models	Latency (s)	Availability	Range (m)	Rate (Gbps)	Application scenarios
Ultra short-range communications	<1	NS	<10	10	Wireless tollgate and kiosks to transfer e-magazine, picture library, 4K movie trailers, 4K movies
Video streaming in smart homes	<0.005	NS	<5	28	8K video stream between a source device (e.g., set-up box, tablet) and a sink device (e.g., smart TV, split TV), replacement of wired interface
Augmented reality	<0.005	NS	<10	20	Interface between a constantly moving high-end wearable devices and its managing device to deliver 3D video
Data center	<0.1	99.99%	<5	40	Inter-rack connectivity, wireless backup connection
Vehicular networks	<0.1	NS	<1000	NS	Intra- and inter-car connectivity, intersection collision avoidance, public safety
Video on-demand	<0.01	NS	<100	NS	Broadcast in crowd public places (e.g., classroom, in flight, train, ship, bus, exhibitions)
Mobile offloading	<0.1	99.99%	<100	20	Offload video traffic from cellular interface to the mmWave interface
Mobile fronthauling	<0.035	99.99%	<200	20	Wireless connections between remote radio heads and base band unit
Mobile backhauling	<0.035	99.99%	<1000	20	Small cell backhauling, mutihop backhauling, inter-building communications

Table 1.
Application scenarios for mmWave networks. This table is deduced from ongoing discussions inside IEEE 802.11ay study group. "NS" means not specified yet.

The commercial potential of mmWave networks initiated several standardization activities within WPANs and wireless local area networks (WLANs), such as IEEE 802.15.3c [6], IEEE 802.11ad [7], WirelessHD consortium, Wireless Gigabit Alliance (WiGig), and recently IEEE 802.11ay study group on next-generation 60 GHz [8].

The unique hardware requirements and propagation characteristics of mmWave networks lead to many challenges at the physical, medium access control (MAC), and routing layers [9]. These challenges are exacerbated by the potential spectrum heterogeneity, i.e., coexistence with the legacy sub-6-GHz systems. Compared to the sub-6-GHz communications, mmWave systems exhibit orders of magnitude higher path loss and atmospheric absorption, higher penetration loss, very sparse-scattering environments, smaller wavelength, a much higher number of antenna elements, which may result in high antenna gains and possible noise-limited operation. These unique features demand a significant reconsideration in the design principles of the communication architecture and protocols, especially at the MAC layer.

2. Fundamentals

2.1 The directed mmWave wireless channel

MmWave communications use frequencies in the range 10–300 GHz. The mmWave systems exhibit high path-loss, high penetration loss, high frequency/short wavelength, and very large bandwidth. The small wavelength allows for the implementation of massive numbers of antennas in both transmitter and receiver, which boosts the achievable antenna gain with some extra signal processing, without affecting the size of their radio chips.

The additional antenna gain can almost completely compensate for the higher path-loss of mmWave communications. A byproduct of the directional communication is the new concept of directional spatial channel, i.e., a channel can be established in a specific direction with a range that varies according to the directionality level [9]. Directional communications and vulnerability to obstacles in mmWave networks have two main consequences: (1) deafness and (2) blockage [9].

Deafness refers to the situation in which a directional communication link cannot be established due to misalignment between the beams of the transmitter and the receiver. To address this problem, we may need a time-consuming procedure of beam training. That is, an operation in which the transmitter and receiver find a beam pair, pointing to each other, which maximizes the link budget. In one hand, the alignment procedure complicates the link establishment phase. On the other hand, it substantially reduces multiuser interference [10], as the receiver listens only to a specific directed mmWave channel. In the extreme case, mmWave networks may operate in a noise-limited regime where multiuser interference is almost completely suppressed and no longer limits the throughput. This is a noticeable change with respect to the conventional interference-limited sub-6 GHz networks. This unique feature makes mmWave suitable for ultra-dense networks (also called massive wireless access) with dense deployments of infrastructure nodes and terminals.

Blockage refers to a high penetration loss due to obstacles that may not be solved by increasing the transmission power. Addressing blockage requires utilizing alternative directed mmWave channels that are not blocked. These channels may be provided by reflectors or intermediate relay nodes. In both solutions, overcoming blockage entails the execution of a new time-consuming beam training procedure.

2.2 Beam training

The use of low-complexity and low-power mmWave devices, along with the massive number of antennas, makes traditional digital beamforming based on instantaneous channel state information very expensive. Instead, the existing standards assume the use of analog beamforming and establish an mmWave link using the so-called beam-searching approach. This approach searches among a set of pre-defined beam steering vectors for the transmitter and the receiver (beam training codebook) and selects the best beam pairs [1, 8, 9]. More specifically, current standards suggest a three-stage beam-searching technique to reduce alignment overhead. After a quasi-omnidirectional sweep with very wide beams, a coarse-grained sector-level sweep is performed, followed by a beam-level refinement phase (the highest resolution pattern specified in the codebook). Each level involves an exhaustive search over all possible transmission and reception directions through a sequence of pilot transmissions. The combination of vectors that maximizes the signal-to-noise ratio (SNR) is then selected for the beamforming.

This beam searching process introduces a new *alignment-throughput* trade-off [11]. That is, on the one hand, a narrower beamwidth enhances the beam resolutions, so increases the alignment overhead and leaves less time for data transmission. On the other hand, it provides a higher antenna gain, leading to a higher transmission rate.

One of the main drawbacks of analog beamforming is the lack of multiplexing gain, which is addressed by the hybrid digital/analog beamforming architecture [12]. Efficient beam training for hybrid beamforming is an active field of research.

2.3 Control channel

Many operations such as establishing a communication channel, discovering neighbors, exchanging routing information, and coordinating channel access rely on the exchange of signaling messages on a control channel. The characteristics of mmWave communications introduce fallback and directionality trade-offs, which also appear in mmWave cellular networks [13].

The *fallback* trade-off is the trade-off between sending control messages through an mmWave or a microwave channel. The mmWave channel is subject to blockage, reducing the reliability of the control channel. A dedicated microwave control channel facilitates network synchronization and broadcasting at the expense of [14]. The cost of this approach is higher hardware complexity and energy consumption since an extra transceiver should be tuned on the microwave control channel. Moreover, a microwave control channel cannot be used to estimate the mmWave channel and adopt proper beamforming. Note that realizing a control channel in the mmWave band with omnidirectional transmission/reception may substantially reduce the system performance. The main reason is a mismatch between the ranges at which a high-quality data link can be established (using directional communications) and the range at which control messages can be exchanged [13].

The *directionality* trade-off arises due to the potential of establishing a control channel with multiple antennas. An omnidirectional control channel is subject to a very short range due to the lack of antenna gains, but it diminishes the deafness problem. A directional control channel benefits longer coverage at the expense of extra alignment overhead.

Altogether, we may have two justifiable realizations for physical control channels: (1) omnidirectional-microwave, employed in ECMA 387 [15], and (2) directional-mmWave, employed in IEEE 802.15.3c [6] and IEEE 802.11ad [7].

3. Initial access and mobility management

Initial access and mobility management are fundamental MAC layer functions that specify how user equipment (UE) should connect to the network and preserve its connectivity. In this section, we highlight important design aspects of initial access that should be considered in mmWave cellular networks.

3.1 Synchronization and cell search

In the long-term evolution (LTE) systems, the so-called primary and secondary synchronization signals enable acquiring time-frequency domain synchronization during the cell search phase. Current cellular networks use beamforming only *after* omnidirectional synchronization and cell search procedure. However, as pointed out in [16], performing cell search on an omnidirectional physical control channel while having antenna gain in data transmission causes a mismatch between the ranges at which a link with reasonable data rate can be established and the range at which a broadcast synchronization signal along with cell identity can be detected. At a normal free-space propagation environment at 28 GHz, the data range can be at least four times larger than the synchronization range with only 30 dBi combined antenna gains. Such a huge mismatch in the ranges of the control and data plane can severely limit the performance of mmWave cellular networks.

3.2 System information

System information includes cell configurations such as frequency band, downlink and uplink bandwidth, cell identity, random access procedure, and the number of transmit and receive antennas. In LTE, the so-called master and system information blocks embed system information. They are transmitted in the physical broadcast dedicated channel and the physical downlink shared channel, respectively. While dedicated control channels can be established with omnidirectional communications, a UE still needs to decode a directional shared channel to extract system information in an mmWave cellular network. Consequently, it will become subject to the fallback and directionality trade-offs. This is a fundamental MAC layer challenge, which is not present in the legacy microwave cellular networks, as all their rendezvous signaling are done in the single antenna mode (omnidirectional physical control channels).

3.3 Random access

At the very beginning, a UE has no reserved channel to communicate with the BS(s). In this case, it sends a channel reservation request using either contention-free or contention-based channel access schemes. In the contention-free approach, the network broadcasts multiple access signals that uniquely poll individual UEs to avoid potential collisions. Upon decoding a signal, each UE knows its uplink parameters including analog or digital beamformer, random access preamble, and allocated resource for transmission of the preamble. Embedding all this information a priori is a challenging task due to the lack of spatial synchronization at the very beginning. The contention-based approach is another strategy to send channel access requests wherein requests may be dropped due to potential collision (in the case of simultaneous transmissions in the same cell) or not be received (in the case of blockage or deafness). The comprehensive analysis of the next chapter shows that small to modest size mmWave networks operating with a simple contention-based protocol (slotted ALOHA) experience a very small collision probability. Moreover, narrower transmission and reception beams reduce the contention level, making contention-based procedures more justifiable than complex and wasteful contention-free ones [17].

In LTE, a UE triggers a timer after sending a contention-based channel access preamble, and upon receiving no response from the base station (BS), it retransmits the preamble after a random waiting (backoff) time or/and with increased transmission power. As we have discussed, the deafness problem of mmWave communications cannot be efficiently addressed by an increased transmission power or a backoff time. In fact, a UE may undergo multiple subsequent backoff executions in the deafness condition, resulting in an unnecessarily prolonged backoff time [18]. To alleviate this problem, [18] introduces a novel collision-notification (CN) signal at the MAC layer to distinguish packet drops due to a collision to those due to deafness and blockage. It builds on the following observation: if the received signal has enough energy but it is not decodable, the receiver declares a collision event. Whereas, if the received signal is not decodable due to lack of energy, the receiver declares a deafness-or-blockage event. During the beam-searching phase, if a BS receives energy from a direction that is not decodable due to collisions, it sends back a CN message in that direction. After transmitting a preamble and depending on the received control signal, a UE will take one of the following actions:

1. A reservation grant is received before the timeout: the UE starts its transmission;

2. A CN message is received before the timeout: the UE assumes a contention in that spatial direction, starts the backoff procedure, and retransmit after a random delay;

3. No signal is received before the timeout: the UE assumes that there is a deafness-or-blockage event in that directed spatial channel, investigates another direction or adjusts the transmission beamwidth.

Action three avoids unnecessary backoff procedures in the case of deafness-or-blockage events, substantially improving the performance of contention-based random access.

3.4 Mobility management and handover

Pencil-beam operations of mmWave systems suppress the interference at the price of more challenging mobility management and handover tasks. Vulnerability to random obstacles, UE mobility, and loss of precise beamforming information may trigger frequent handovers if only the received signal strength indicator (RSSI) is used as a reassociation metric [13]. Every handover may entail a beam-training overhead. In the presence of frequent handovers, the transmitter and receiver may remain most of the time in the beam-training phase rather than the data transmission phase.

To alleviate extra handovers due to random blockage, we can adopt the following association options in mmWave networks:

- Multiple parallel connectivity, and

- Single sequential connectivity.

In the first approach, a client adopts multi-beam transmissions toward several BSs (relays) at the same time to establish multiple paths, either at the same band [13] or different frequency bands [19]. This approach provides seamless handover, continuous connectivity, and blockage robustness. However, we may observe an SNR loss for each beam (when a transmitter uses a fixed total power budget), higher signaling and computational complexities for beamforming, and more complicated resource management and relay selection. Cell-free access is a very similar approach wherein a processor unit coordinates the communication among multiple BSs and UEs, without associating a UE to a particular BS [20]. To enable this mode, we may need to have digital or hybrid beamforming [21]. To reduce the computational and signaling overhead of the beamforming with many antenna elements, current mmWave standards adopt analog beamforming, avoiding the realization of multiple parallel connectivity. Instead, a client may be associated with several BSs (relays) with several paths and establishes a data channel using only one of these paths, while using the others as backups. This single sequential connectivity scenario, as reported in [22], is standard-compliant and mitigates disadvantages of the multiple parallel connectivity scenario. Moreover, recent works exploited sparse scattering properties of mmWave channels to model the spatial and temporal correlation of the mmWave channels between a stationary BS and a mobile UE [23, 24]. Using this model, they have proposed efficient beam-tracking approaches to predict and facilitate handover events.

4. Resource allocation

4.1 Interference characteristics

To design a proper hybrid MAC for mmWave networks, the main steps are analyzing the multiuser interference, evaluating performance gain (in terms of throughput/delay) due to various resource allocation protocols, and investigating the signaling and computational complexities of those protocols. Roughly speaking, as the system goes to the noise-limited regime, the required complexity for proper resource allocation and interference avoidance functions at the MAC layer substantially reduces [11, 25, 26]. For instance, in a noise-limited regime, a very simple resource allocation such as activating all links at the same time without any coordination among different links may outperform a complicated independent-set based resource allocation [11]. Instead, pencil-beam operation complicates negotiation among different devices in a network, as control message exchange may require time-consuming antenna alignment (beam-training) procedure to avoid deafness.

The seminal work in [10] shows the existence of a noise-limited regime (also called pseudowired abstraction) in outdoor mmWave mesh networks. However, indoor mmWave WPANs may not be noise-limited, as shown in [11, 25–27]. In particular, the optimal resource allocation policy may need to deactivate some links to handle the non-negligible multiuser interference [11]; the noise power is not always the limiting factor. To have a concrete example, we have simulated an ad hoc network with a random number of mmWave links deployed on a 10×10 m^2 area, all operating with the same beamwidth at 60 GHz. Each transmitter/receiver is aligned to its own communication link, and they are active with some probability p independent of the activity of the other links. We assume 2.5 mW transmission power, 16 dB/Km atmospheric absorption, (on average) one obstacle on every a 4 m^2 area, and sector blockage model [28]. We computed and depicted in **Figure 1** the area spectral efficiency (ASE), defined as the network sum-rate divided by the area size,

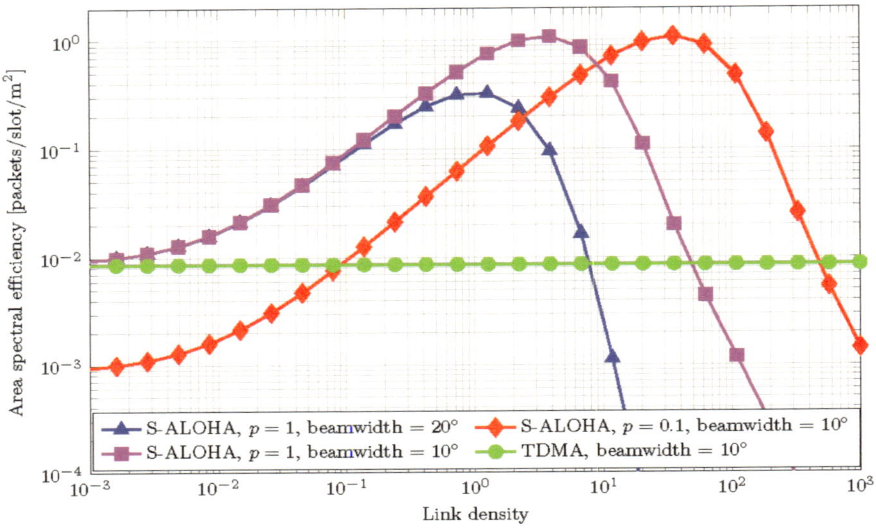

Figure 1.
Area spectral efficiency against link (one transmitter-receiver pair) density (m^2). Area size is 10×10 m^2. The obstacle density is one every 4 m^2. Slotted ALOHA provides substantially higher area spectral efficiency, compared to TDMA. These performance gains may improve with the number of links.

as the performance measure. From this figure, increasing the number of links in the network does not affect ASE of TDMA, which is slightly lower than one packet per slot. The reason is packet loss due to blockage on some links, and this loss almost vanishes when the obstacle density goes to zero. Slotted ALOHA with transmission probability $p = 1$ provides the highest ASE, which is firstly increasing with the link density and then shows a strictly decreasing behavior due to excessive collision. Using narrower a beamwidth or lower transmission probability alleviates the collision level and improves the ASE at the expense of the alignment-throughput trade-off [11].

The numbers of the figure indicate that, from the perspective of ASE, mmWave networks benefit from dense deployment, yet every link may observe some level of performance degradation when we increase the number of devices in the network [11]. Reference [29] reported similar observations in mmWave cellular networks. Such performance drops imply that the accuracy of the noise-limited assumption to model the actual network behavior reduces with the number of links. The increased directionality level in a mmWave network reduces multiuser interference; however, this reduction may not be enough to take an action (e.g., resource allocation) based on the assumption of being in a noise-limited regime. Consequently, a pseudowired assumption may be detrimental for proper MAC layer design. However, the interference footprint may not be so large that we need to adopt very conservative resource allocation protocols such as time division multiple access (TDMA), which activates only one link at a time, as already adopted by the current mmWave standards [6, 7]. Reference [30] proposed an index to quantify the impact of various components of the interference models and to propose a tractable and accurate interference model for mmWave networks.

4.2 Beam-searching and concurrent transmission

Despite the small interference footprint of mmWave networks, the option of concurrent transmissions scheduling was not included in the existing standards and proposed only recently. The authors of [27] consider the problem of maximizing the number of scheduled flows such that their quality of service requirement is not violated. A greedy scheduling scheme is proposed, where an additional link is activated in each time slot if it improves the total throughput, i.e., throughput gain from this extra link overweighs the performance drop due to additional interference. Reference [31] proposed a similar greedy heuristic by designing a priority ordering of the links. As long as the signal to interference plus noise ratio at all receivers exceeds a threshold, additional links are activated according to this list. The main issue of those approaches is that they are reactive protocols, i.e., a link has to be activated to deduce if it is compatible with other transmissions.

In proactive protocols, one may need to address the alignment-throughput trade-off while designing concurrent scheduling problem. Because, a narrow beamwidth, besides boosting the link budget, reduces multiuser interference, so increases SINR and thereby achievable transmission rate. However, the price of this rate enhancement is higher alignment overhead per-link and possibly complicated scheduling procedures. Reference [11] addressed this problem by an optimization problem that brings together beam-searching and transmission scheduling using estimates of the interference terms.

4.3 Association

Association governs the long-term allocation of communication resources among various BSs. Due to high penetration loss, blockage in mmWave networks may be only addressed by re-association or relaying procedures not by increasing the transmit power [9].[1] The association and relaying are particularly important in mmWave networks due to the dense deployment of the BSs and limited size of the cells [32]. Relaying techniques can provide a more uniform quality of service by offering robust mmWave connection, load balancing, coverage extension, indoor-outdoor coverage, efficient mobility management, and smooth handover operation [1, 9, 13, 32–34]. As shown in, an alternative path through relay nodes can increase the connectivity of a mmWave system by about 100%. Furthermore, the relaying technique can enable high-quality live video streaming over 300 m [34]. Therefore, proper association of clients to BSs and relaying techniques are very important routines in mmWave networks.

Developing proper association techniques has been the focus of intense research in the last years [35–40], as it may govern the long-term resource allocation policies of conventional wireless networks [35]. The current mmWave standards use the minimum-distance association, which leads to a simple association metric based on the RSSI [14]. This metric is proved to be suitable for an interference-limited homogenous network, but it may lead to poor use of the available resources in the presence of a non-uniform spatial distribution of clients, non-interference-limited environments, and heterogeneous BSs/relays with a different number of antenna elements and different transmission powers [35]. It may lead to an unbalanced number of clients per BS, drastically reducing the available resources per client in highly populated areas [13] while wasting resources in sparse areas. This poor load balancing indeed decreases network-wide fairness, since overloaded BSs cannot provide their associated clients as much resource as less-loaded BSs. Thus, it is possible for the clients to associate with farther BSs for better load sharing.

Besides the existing association techniques of the current mmWave standards, there are many more solutions for the association and relaying from the literature of microwave networks. In [37], a client association policy is investigated to ensure

[1] As an alternative approach, link establishment via reflectors may address a blockage, given the existence of such reflectors with sufficiently large reflection indices.

network-wide max-min fair bandwidth allocation to the clients in WLANs. In the seminal work of [36], a joint association and resource allocation problem is formulated for a heterogeneous cellular network to ensure network-wide fairness, by a distributed solution algorithm. These association procedures are highly sub-optimal for mmWave networks due to frequent handovers of mmWave networks and small interference footprint. Reducing the overhead of frequent reassociation, together with the natural need of load balancing among the BSs, justifies that a client in mmWave networks may be advantageously served by a farther but less-loaded and easy-to-find BS [13]. Robustness of the association to random blockage should be improved to reduce the number, and thereby the overhead/delay, of reassociation and to provide a seamless handover [9, 13]. Reference [41] addressed the association problem in 60 GHz mmWave communications. However, it did not consider relays, a vital part of mmWave networks, which substantially increases the difficulty of the association and relaying problem. This problem has been addressed in [42] where the authors showed that the optimal relay selection improves the load-balancing throughout the network and affects heavily the ability of a terminal to reach a farther BS. Moreover, [22] proposed an adaptive reassociation mechanism for time-varying mmWave networks, wherein the previous association solution is used as a proper initial guess to solve a new network-wide association optimization problem.

4.4 Spectrum sharing

Spectrum sharing between multiple operators was recently proposed as a way to allow more efficient use of the spectrum in mmWave networks. Preliminary studies have shown that the specific features of mmWave frequencies, including the propagation characteristics and narrow beam operations, facilitate spectrum sharing in the mmWave bands. Reference [43] proposed a mechanism to let two different IEEE 802.11ad access points transmit over the same time/frequency resources. To realize this mechanism, the authors introduced a new signaling report, which is broadcast by each access point to establish an interference database that facilitates scheduling decisions. A similar approach was proposed in [44] for mmWave cellular systems, with both centralized and distributed coordination among operators. In the centralized case, a new architectural entity determines the links that cannot be concurrently activated, based on the reports of the interference powers. In the decentralized case, the victim network sends a message to the interfering network. The two networks can further refine the coordination pattern via multiple iterations.

Reference [45] investigated the feasibility of sharing the mmWave spectrum between the device-to-device/cellular and access/backhaul networks and proposed a new MAC layer in order to regulate concurrent transmissions in a centralized manner. Given the sporadic presence of strong interference in mmWave networks, reference [13] showed the need for only on-demand inter-cell interference coordination as opposed to often heavy coordination requirements of spectrum sharing at the sub-6-GHz bands. Reference [46] investigated the feasibility of spectrum sharing in mmWave cellular networks and showed that, under certain conditions such as idealized antenna pattern, spectrum sharing may be beneficial even without any coordination in the entire network. Reference [47] showed that infrastructure sharing in mmWave cellular networks is also beneficial and its gain is almost identical to that of spectrum sharing. Reference [48] discussed the architectures and protocols required to make spectrum sharing work in practical mmWave cellular networks and provided preliminary results regarding the importance of coordination. Reference [49] studied the performance of a hybrid spectrum scheme in which exclusive access is used at frequencies in the 20–30 GHz range while spectrum sharing (or even unlicensed spectrum) is used at frequencies around 70 GHz.

5. Conclusions

This chapter summarized main characteristics of mmWave systems, including severe attenuation, sparse-scattering environment, huge bandwidth, blockage and deafness, and possible noise-limited operation. We discussed initial access and mobility management (e.g., synchronization, random access, and handover), characterized interference footprint and reviewed existing solutions for resource allocation in mmWave networks.

Acknowledgements

This work was supported in part by the Swedish Research Council project 2018-00820.

Author details

Hossein S. Ghadikolaei
KTH Royal Institute of Technology, Stockholm, Sweden

*Address all correspondence to: hshokri@kth.se

References

[1] Niu Y, Li Y, Jin D, Su L, Vasilakos A. A survey of millimeter wave communications mmWave for 5G: Opportunities and challenges. Wireless Networks. Nov 2015;**21**(8):2657-2676

[2] Rappaport TS et al. Millimeter wave mobile communications for 5G cellular: It will work! IEEE Access. 2013;**1**:335-349

[3] Andrews JG et al. What will 5G be? IEEE Journal on Selected Areas in Communications. 2014;**32**(6):1065-1082

[4] Rappaport TS, MacCartney GR, Samimi MK, Sun S. Wideband millimeter-wave propagation measurements and channel models for future wireless communication system design. IEEE Transactions on Communications. 2015;**63**(9):3029-3056

[5] Shokri-Ghadikolaei H. Millimeter-wave Networking: Fundamental Limits, Scalable Algorithms, and Design Insights. Stockholm, Sweden: KTH Royal Institute of Technology; 2017

[6] IEEE 802.15.3c P art 15.3: Wireless medium access control (MAC) and physical layer (PHY) specifications for high rate wireless personal area networks (WPANs) amendment 2: Millimeter-wave-based alternative physical layer extension. 2009

[7] IEEE 802.11ad. P art 11: Wireless LAN medium access control (MAC) and physical layer (PHY) specifications—Amendment 3: Enhancements for very high throughput in the 60 GHz band. 2012

[8] Zhou P, Cheng K, Han X, Fang X, Fang Y, He R, et al. IEEE 802.11 ay-based mmWave WLANs: Design challenges and solutions. IEEE Communication Surveys and Tutorials. 2018;**20**(3):1654-1681

[9] Rappaport TS, Heath R, Daniels RC, Murdock JN. Millimeter Wave Wireless Communications. New Jersey, United States: Prentice Hall, Pearson Education; 2014

[10] Singh S, Mudumbai R, Madhow U. Interference analysis for highly directional 60-GHz mesh networks: The case for rethinking medium access control. IEEE/ACM Transactions on Networking. 2011;**19**(5):1513-1527

[11] Ghadikolaei HS, Gkatzikis L, Fischione C. Beam-searching and transmission scheduling in millimeter wave communications. In: Proceedings IEEE International Conference on Communications (ICC); 2015. pp. 1292-1297

[12] Kutty S, Sen D. Beamforming for millimeter wave communications: An inclusive survey. IEEE Communication Surveys and Tutorials. 2016;**18**(2):949-973

[13] Ghadikolaei HS, Fischione C, Fodor G, Popovski P, Zorzi M. Millimeter wave cellular networks: A MAC layer perspective. IEEE Transactions on Communications. 2015;**63**(10):3437-3458

[14] Nitsche T, Flores AB, Knightly EW, Widmer J. Steering with eyes closed: Mm-Wave beam steering without in-band measurement. In: Proceedings—IEEE INFOCOM; 2015. Vol. 26; pp. 2416-2424

[15] ECMA-TC48, ECMA standard 387, High rate 60 GHz PHY, MAC and HDMI PAL. 2008

[16] Li QC, Niu H, Wu G, Hu RQ. Anchor-booster based heterogeneous networks with mmwave capable booster cells. In: Proceedings IEEE Globecom Workshops; 2013. pp. 93-98

[17] Jeong C, Park J, Yu H. Random access in millimeter-wave beamforming cellular networks: Issues and approaches. IEEE Communications Magazine. 2015;**53**(1):180-185

[18] Ghadikolaei HS, Fischione C, Popovski P, Zorzi M. Design aspects of short range millimeter wave wireless networks: AMAC layer perspective. In: IEEE Networks; 2015

[19] Polese M, Giordani M, Mezzavilla M, Rangan S, Zorzi M. Improved handover through dual connectivity in 5G mmWave mobile networks. IEEE Journal on Selected Areas in Communications. 2017;**35**(9):2069-2084

[20] Ngo HQ, Ashikhmin A, Yang H, Larsson EG, Marzetta TL. Cell-free massive mimo versus small cells. IEEE Transactions on Wireless Communications. 2017;**16**(3):1834-1850

[21] Femenias G, Riera-Palou F. Cell-free millimeter-wave massive mimo systems with limited fronthaul capacity. IEEE Access. 2019;**7**:44596-44612

[22] Xu Y, Ghadikolaei HS, Fischione C. Adaptive distributed association in time-variant millimeter wave networks. IEEE Transactions on Wireless Communications. 2019;**18**(1):459-472

[23] Zhou A, Zhang X, Ma H. Beam-forecast: Facilitating mobile 60 GHz networks via model-driven beam steering. In: Proceedings—IEEE INFOCOM; 2017

[24] Zhou A, Wu L, Xu S, Ma H, Wei T, Zhang X. Following the shadow: Agile 3-D beam-steering for 60 GHz wireless networks. In: Proceedings—IEEE INFOCOM; April 2018; 2018. pp. 2375-2383

[25] Qiao J, Shen X, Mark J, Shen Q, He Y, Lei L. Enabling device-to-device communications in millimeter-wave (5G) cellular networks. IEEE Communications Magazine. 2015;**53**(1):209-215

[26] Park M, Gopalakrishnan P. Analysis on spatial reuse and interference in 60-GHz wireless networks. IEEE Journal on Selected Areas in Communications. 2009;**27**(8):1443-1452

[27] Qiao J, Cai LX, Shen X, Mark J. STDMA-based scheduling algorithm for concurrent transmissions in directional millimeter wave networks. In: Proceedings IEEE International Conference on Communications (ICC); 2012. pp. 5221-5225

[28] Ghadikolaei HS, Fischione C. The transitional behavior of interference in millimeter wave networks and its impact on medium access control. IEEE Transactions on Communications. 2016;**62**(2):723-740

[29] Di Renzo M. Stochastic geometry modeling and analysis of multi-tier millimeter wave cellular networks. IEEE Transactions on Wireless Communications. 2015;**14**(9):5038-5057

[30] Ghadikolaei HS, Fischione C, Modiano E. Interference model similarity index and its applications to mmWave networks. IEEE Transactions on Wireless Communications. 2018;**17**(1):71-85

[31] Wu X et al. FlashLin Q: A synchronous distributed scheduler for peer-to-peer ad hoc networks. IEEE/ACM Transactions on Networking. 2013;**21**(4):1215-1228

[32] Singh S, Ziliotto F, Madhow U, Belding EM, Rodwell M. Blockage and directivity in 60 GHz wireless personal area networks: From cross-layer model to multihop MAC design. IEEE Journal on Selected Areas in Communications. 2009;**27**(8):1400-1413

[33] Rangan S, Rappaport TS, Erkip E. Millimeter wave cellular wireless networks: Potentials and challenges. Proceedings of the IEEE. 2014;**102**(3):366-385

[34] Kim J, Tian Y, Mangold S, Molisch AF. Joint scalable coding and routing for 60 GHz real-time live HD video streaming applications. IEEE Transactions on Broadcasting. 2013;**59**(3):500-512

[35] Andrews JG, Singh S, Ye Q, Lin X, Dhillon HS. An overview of load balancing in HetNets: Old myths and open problems. IEEE Wireless Communications. 2014;**21**(2):18-25

[36] Ye Q, Rong B, Chen Y, Al-Shalash M, Caramanis C, Andrews JG. User association for load balancing in heterogeneous cellular networks. IEEE Transactions on Wireless Communications. 2013;**12**(6):2706-2716

[37] Bejerano Y, Han S. Fairness and load balancing in wireless LANs using association control. IEEE/ACM Transactions on Networking. June 2007;**15**(3):560-573

[38] Hui J, Devetsikiotis M. A unified model for the performance analysis of IEEE 802.11e EDCA. IEEE Transactions on Communications. 2005;**53**(9):1498-1510

[39] Boostanimehr H, Bhargava VK. Unified and distributed QoS-driven cell association algorithms in heterogeneous networks. IEEE Transactions on Wireless Communications. 2015;**14**(3):1650-1662

[40] Shokri-Ghadikolaei H, Boccardi F, Fischione C, Fodor G, Zorzi M. Spectrum sharing in mmWave cellular networks via cell association, coordination, and beamforming. IEEE Journal on Selected Areas in Communications. 2016;**34**(11):2902-2917

[41] Athanasiou G, Weeraddana C, Fischione C. Auction-based resource allocation in millimeter-wave wireless access networks. IEEE Communications Letters. 2013;**17**(11):2108-2111

[42] Xu Y, Shokri-Ghadikolaei H, Fischione C. Distributed association and relaying with fairness in millimeter wave networks. IEEE Transactions on Wireless Communications. 2016;**15**(12):7955-7970

[43] Feng W, Li Y, Jin D, Zeng L. Inter-network spatial sharing with interference mitigation based on IEEE 802.11ad WLAN system. In: Proceedings IEEE Global Communications Conference (GLOBECOM) Workshop; 2014. pp. 725-758

[44] Li G, Irnich T, Shi C. Coordination context-based spectrum sharing for 5G millimeter-wave networks. In: Proceedings International Conference on Cognitive Radio Oriented Wireless Networks and Communications (CROWNCOM); 2014. pp. 32-38

[45] Niu Y, Gao C, Li Y, Su L, Jin D, Vasilakos A. Exploiting device-to-device communications in joint scheduling of access and backhaul for mmWave small cells. IEEE Journal on Selected Areas in Communications. 2015;**33**(10):2052-2069

[46] Gupta AK, Andrews JG, Heath RW. On the feasibility of sharing spectrum licenses in mmWave cellular systems. IEEE Transactions on Communications. 2016;**64**(9):3981-3995

[47] Rebato M, Mezzavilla M, Rangan S, Zorzi M. Resource sharing in 5G millimeter-wave bands. In: Proceedings IEEE International Conference on Computer Communications (INFOCOM) Workshop; 2016

[48] Boccardi F et al. Spectrum pooling in mmWave networks:

Opportunities, challenges, and enablers. IEEE Communications Magazine. 2016;**54**(11):33-39

[49] Rebato M, Boccardi F, Mezzavilla M, Rangan S, Zorzi M. Hybrid spectrum access for mmWave networks. In: Proceedings IEEE IFIP Annual Mediterranean Ad Hoc Networking (MedHocNet) Workshop; 2016

Chapter 8

Monte Carlo Radiative Transfer Modeling of Underwater Channel

Rafael M.G. Kraemer, Luís M. Pessoa and Henrique M. Salgado

Abstract

The radiative transfer equation (RTE) is a theoretical framework that can be used for predicting and interpreting underwater light fields in terms of the constituents of natural water bodies. However, the RTE is a complex integrodifferential equation and deriving exact solutions for it is a difficult task. In this chapter, we aim to present some details regarding Monte Carlo simulations and how this method may be applied to solve the RTE numerically. By solving the RTE, one may accurately predict the received power and estimate the channel bandwidth and several other measurable parameters with regard to multiple water conditions. Simulations will also be presented.

Keywords: RTE, underwater, optical, wireless, channel, propagation, photons, UOWC

1. Introduction

When compared with free space optical (FSO) communication channel, the underwater optical wireless communication (UOWC) shows unique characteristics. Because of this, the channel models usually used in FSO are not suited for UOWC, and different channel models must be developed to describe the different photon interactions under play.

Mobley [1] classified the optical properties of water into two different groups, inherent optical properties (IOPs) and apparent optical properties (AOPs). The two major IOPs that will attenuate light propagation in UOWC are absorption and scattering. Hence, any attempt in modeling the UOWC channel will need to consider absorption and scattering effects within specific link configurations.

Several theoretical models were employed by researchers to model the UOWC channel, being the Beer-Lambert law the most employed due to its simplicity [2]. Despite its simplicity, the Beer-Lambert law contains two implicit assumptions: that the transmitter and receiver are perfectly aligned and that all scattered photons are lost. The assumption that all photons that undergo scattering are lost will severely underestimate the received optical power, especially in water environments where scattering is the dominant IOP.

Another theoretical model for modeling the underwater channel is the radiative transfer equation (RTE) [3]. The RTE simply says that as a beam of radiation travels through a medium, in this case water, it will lose energy to absorption, gain energy by emission, and redistribute energy by scattering. Yet, as the RTE is an integrodifferential equation involving different independent variables, an exact analytical solution is hard to find. In view of this, solving the RTE numerically is a preferred approach, being the most popular one the Monte Carlo simulation.

2. The radiative transfer equation

One important law of geometrical radiometry is the n-squared law for radiance. To derive it, we consider two mediums separated by a transparent surface; in that case, Snell's law states that the angles of an incident ray from medium 1 to medium 2 are related by:

$$n_1 \sin(\theta_1) = n_2 \sin(\theta_2), \tag{1}$$

where n is the index of refraction of the medium. Considering that a ray is simply a narrow beam of photons traveling in almost the same direction, and if we consider two narrow rays, incident and refracted, that will have solid angles given by:

$$\Delta\Omega_1 = \sin\theta_1 \Delta\theta_1 \Delta\phi, \tag{2}$$

$$\Delta\Omega_2 = \sin\theta_2 \Delta\theta_2 \Delta\phi, \tag{3}$$

and as seen in [3], squaring each side of Eq. (1) and multiplying by the Azimuthal spread, $\Delta\phi$, we obtain:

$$n_1^2 \cos\theta_1 \Omega_1 = n_2^2 \cos\theta_2 \Omega_2, \tag{4}$$

which is known as the Straubel's invariant. Consider now the radiances of the two rays, incident and refracted, defined as:

$$L_1 = \frac{\Delta P_1}{\Delta A_1 \Delta\Omega_2}, \tag{5}$$

$$L_2 = \frac{\Delta P_2}{\Delta A_2 \Delta\Omega_2}, \tag{6}$$

where $\Delta P_{1,2}$ is the spectral radiant power and $\Delta A_{1,2}$ is the cross section area of incident and refracted rays. The ratio between the refracted and incident rays is called the Fresnel transmittance:

$$\frac{\Delta P_1}{\Delta P_2} = T, \tag{7}$$

from which we can obtain:

$$\frac{L_2}{n_2^2} = T\frac{L_1}{n_1^2}, \tag{8}$$

which is the n-squared law for radiance. For the case of $T = 1$, which is the case of normal incidence on an air-water surface, Eq. (8) becomes:

$$\frac{L_2}{n_2^2} = \frac{L_1}{n_1^2}. \tag{9}$$

As seen in [3], this result is known as the fundamental theorem of radiometry and it states that the radiance divided by the refraction index squared is constant along any path; however, because all real substances will cause some absorption and scattering on the incident photons, the validity of theorem is restricted for paths in vacuum.

If we consider that a ray of radiance L_1 starts in a medium with index of refraction n_1 and goes through successive refractions and ends up with radiance L_S in a medium of refractive index n_s, we can come up with a formulation for successive crossings of a ray from one medium to another:

$$\frac{L_S}{L_1} = \prod_{j=1}^{s-1} T(j,j+1)\frac{n_s^2}{n_1^2}, \tag{10}$$

where $T(j,j+1)$ is the Fresnel transmittance of the interface between the media with refractive indexes n_j and n_{j+1}. From this, we can see that the radiance along a path will change due to variations in the real index of refraction along that same path. The index of refraction in a water body will change from point to point by random molecular motions, by organics or inorganic particulate matter and by turbulent fluctuations in temperature and salinity.

Considering a beam of radiance traveling from point $\vec{x}$ to point $\vec{x} + \Delta \vec{x}$ and taking the limit as $\Delta r = \left|\Delta \vec{x}\right| \to 0$:

$$\lim_{\Delta r \to 0} \frac{\Delta(L/n^2)}{\Delta r} = \frac{D(L/n^2)}{Dr}, \tag{11}$$

being D/Dr the total rate of change along the path. The total rate of change can be expressed in terms of the advective derivative, or the substantive derivative, as referred by Mobley:

$$\frac{D}{Dr} = \frac{1}{v}\frac{D}{Dt}, \tag{12}$$

where v is the speed of light in the medium at position $\vec{x}$ at time t for wavelength λ:

$$\mathrm{v}\left(\vec{x};t;\lambda\right) = \mathrm{c}/\mathrm{n}\left(\vec{x};t;\lambda\right), \tag{13}$$

The operator D/Dt is:

$$\frac{D}{Dt} = \frac{\partial}{\partial t} + \vec{v} \cdot \nabla, \tag{14}$$

and the gradient operator ∇ is defined in the usual way.

From these developments, we can rewrite Eq. (11) as:

$$\frac{D}{Dr} = \frac{1}{v}\frac{D}{Dt} = \frac{1}{v}\frac{\partial}{\partial t} + \hat{\xi} \cdot \nabla, \tag{15}$$

valid for photons traveling with speed v in the direction $\hat{\xi} = \vec{v}/v$.

Writing the expression for the change in L/n^2 along a path that combines all the physical phenomena causing that change, we obtain the most general form of the RTE for unpolarized radiance:

$$\frac{1}{v}\frac{\partial}{\partial t}\left(\frac{L}{n^2}\right)+\hat{\xi}\cdot\nabla\left(\frac{L}{n^2}\right)=-c\left(\frac{L}{n^2}\right)+L^E+L^I+L^S. \tag{16}$$

In Eq. (17), c is the beam attenuation coefficient, $L^E, L^I,$ and L^S are the path functions for elastic scattering, inelastic scattering, and spontaneous emission, respectively. The path function for spontaneous emission is also referred in [3] as the source path function and can be considered as the contribution to total radiance from a light source and in the case of UOWC systems a laser diode or a light emitting diode (LED).

For UOWC systems, we are interested in the time-independent RTE in horizontally homogenous water bodies with a constant index of refraction. Because of this, the factor n^{-2} will divide both sides of Eq. (15). Hence, with $\partial L/\partial t=0$ and using the notation adopted by Mobley where $x_3=z$ and $\xi_3=\cos\theta$, we obtain:

$$\hat{\xi}\cdot\nabla L=\xi_3\frac{\partial L}{\partial x_3}=\cos\theta\frac{dL(\theta,\phi)}{dz}. \tag{17}$$

It is also usual [3] to combine L^I and L^S as an effective source function $S=L^I+L^S$. The RTE then becomes:

$$\cos\theta\frac{dL(\theta,\phi)}{dz}=-cL(\theta,\phi)+L^E(\theta,\phi)+S(\theta,\phi), \tag{18}$$

and the path function for elastic scattering, L^E:

$$L^E(\theta,\phi)=\int_{4\pi}L(\theta',\phi')\beta(\theta',\phi'\rightarrow\theta,\phi)d\Omega(\theta',\phi'), \tag{19}$$

Combining all these terms in Eq. (18), we can come up with the standard form of the RTE. As a simple statement, the IOPs and boundary conditions will go into the RTE and the final radiance will come out as a result. The standard form of the RTE, as shown in [4], is given by:

$$\cos\theta\frac{dL(\theta,\phi)}{cdz}=-\mathrm{L}(\theta,\phi)+\omega_0\int_{4\pi}\widetilde{\beta}(\theta',\phi'\rightarrow\theta,\phi)L(\theta',\phi')d\Omega(\theta',,\phi')+S(\theta,\phi), \tag{20}$$

where $L(\theta,\phi)$ represents the radiance in the direction (θ,ϕ), ω_0 is the single scattering albedo, c is the beam attenuation coefficient, and $\widetilde{\beta}(\theta,\phi)$ is the scattering phase function. Each one of these terms will be explained in more detail in the following sections.

2.1 Inherent optical properties

During the interaction of a photon with a water molecule, one of two things may happen: the photon may be absorbed, leaving the water molecule in a state with higher internal energy, or the photon may undergo scattering. Such scattering occurs when there is a change in the direction of the propagation of the photon or in the photon energy—or in both.

When scattering happens, if the molecule that interacted with the photon returns to its original internal energy state by emitting a photon of the same energy as the absorbed one, the scattering process is called elastic scattering; on the other hand, if the excited molecule emits a photon of lower energy, i.e., a photon with a longer wavelength, the incident photon goes through a process called inelastic scattering. As seen in [3], the absorption coefficient, that will be introduced later in this chapter, accounts for both the conversion of radiant energy into heat and the loss of power at wavelength λ by inelastic scattering to other wavelengths. Mobley [3] refers to this as true absorption. In view of this, in the remaining of this chapter, we will treat the scattering as elastic, the loss of energy due inelastic scattering being included in the absorption coefficient.

To derive a mathematical coefficient for scattering and absorption, we assume the model shown in **Figure 1**.

A volume of water ΔV with thickness Δr is illuminated by a light beam with wavelength λ. Some part of the incident power, $P_I(\lambda)$, will be absorbed within the volume of water, $P_A(\lambda)$; some part, $P_S(\lambda, \theta)$, is scattered out of the beam at an angle θ and the remaining light power, $P_T(\lambda)$, is transmitted. Then by conservation of energy:

$$P_I(\lambda) = P_A(\lambda) + P_S(\lambda, \theta) + P_T(\lambda). \tag{21}$$

Using Eq. (22), we can define the ratio between the absorbed and incident power as the absorptance and the ratio between scattered power and incident power as scatterance. As already stated, when solving the RTE the IOPs usually employed are the absorption and the scattering coefficients which are the absorptance and scatterance per unit distance of the medium. Taking the limit of these terms as the water thickness Δr becomes infinitesimally small, we have:

$$a(\lambda) = \lim_{\Delta r \to 0} \frac{P_A(\lambda)}{P_I(\lambda)\Delta r}. \tag{22}$$

$$b(\lambda) = \lim_{\Delta r \to 0} \frac{P_S(\lambda)}{P_I(\lambda)\Delta r}. \tag{23}$$

The beam attenuation coefficient present in the RTE (Eqs. (16) and (20)) is written as:

$$c(\lambda) = a(\lambda) + b(\lambda). \tag{24}$$

Another useful IOP showing in the RTE is the single scattering albedo, ω_0, that is defined as the ratio between the scattering and the beam attenuation coefficient:

$$\omega_0 = \frac{b(\lambda)}{c(\lambda)}. \tag{25}$$

In waters where the beam attenuation is due primarily to scattering, ω_0 is near one and when the beam attenuation is due primarily to absorption, ω_0 is near zero.

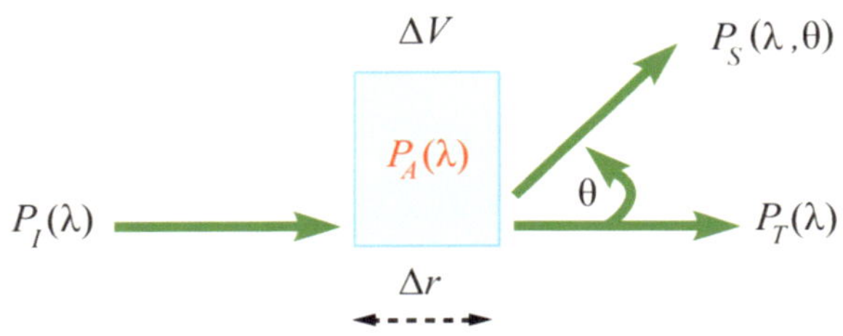

Figure 1.
Geometry of IOPs for a volume of water ΔV.

In [3], it is possible to find the absorption, scattering, and the attenuation coefficients for several water types and wavelengths. The usual parameters considered for UOWC systems are the ones presented in **Table 1**.

As mentioned before, the ratio between the scattered power and incident power is referred as scatterance, that is the fraction of incident power scattered out of the beam through an angle θ into a solid angle $\Delta\Omega$. Defining the angular scatterance per unit distance and unit solid angle, $\beta(\theta, \lambda)$, as:

$$\beta(\theta,\lambda) = \lim_{\Delta r \to 0} \lim_{\Delta\Omega \to 0} \frac{P_S(\theta,\lambda)}{P_I(\lambda)\Delta r \Delta\Omega}. \tag{26}$$

In [3], we can see that the spectral power scattered into the given solid angle $\Delta\Omega$ is just as the spectral radiant intensity, $I_S(\theta, \lambda)$, scattered into the angle θ times the solid angle:

$$P_S(\theta,\lambda) = I_S(\theta,\lambda)\Delta\Omega, \tag{27}$$

if the incident power falls on an area ΔA, then the corresponding irradiance $E_i(\lambda) = P_I(\lambda)/\Delta A$. Recalling that ΔV is the volume of water that is illuminated by the incident beam:

$$\beta(\theta,\lambda) = \lim_{\Delta V \to 0} \frac{I_S(\theta,\lambda)}{E_i(\lambda)\Delta V}. \tag{28}$$

As said in [5], the form of Eq. (28) suggests the name volume scattering function (VSF), that is, the scattered intensity per unit incident irradiance per unit volume of water. Integrating $\beta(\theta, \lambda)$ over all directions gives the total scattered power per unit of incident irradiance and unit volume of water, or as defined in Eq. (23), the scattering coefficient:

$$b(\lambda) = 2\pi \int_0^{\pi} \beta(\theta,\lambda)\sin\theta d\theta. \tag{29}$$

Normalizing Eq. (26) with the scattering coefficient, we obtain the scattering phase function (SPF):

$$\tilde{\beta}(\theta,\lambda) = \frac{\beta(\lambda,\theta)}{b(\lambda)}. \tag{30}$$

The scattering phase function can be interpreted as a probability density function (PDF) for scattering from an incident direction (θ', ϕ') to a final direction (θ, ϕ) [5]. In general, the SPF must be solved numerically, or using tabulated data like the ones given by Petzold [6]. The scattering measurements made by Petzold

Water type	$c(\lambda)\ m^{-1}$	ω_0
Clear waters	0.15	0.25
Coastal waters	0.4	0.55
Harbor I waters	1.1	0.83
Harbor II waters	2.19	0.83

Table 1.
Measured parameters for different water types.

are the most precise to this date, to the best of the author's knowledge, and widely cited in the literature. The measurements were carried in the Bahamas, San Pedro in California, and San Diego harbor also in California, corresponding to clear, coastal, and harbor waters, respectively.

3. Solving the radiative transfer equation: Monte Carlo simulation

The Monte Carlo name was coined by Nicholas Metropolis [7]. The modern version of the method was invented in the late 1940s and found use early use in the studies of neutron transport for the design of nuclear weapons [8, 9]. More recently, Monte Carlo techniques evolved and are now used to solve problems in physical sciences, economics, engineering, and pure mathematics.

Monte Carlo methods may be used to solve any problem that have a probabilistic interpretation, meaning that if the probability of occurrence of each separate event in a sequence of events is known, one can determine the probability that the entire sequence of events will occur. By tracing the fate of millions of photons according to statistical probabilities, a solution for the RTE is built. Each simulated photon path is randomly distinct from the others, as determined by the probabilities of absorption and scattering in the underwater channel.

The Monte Carlo simulations of photon propagation offer a flexible yet rigorous approach toward solving numerically the RTE. In the last decades, the exponential improvement in computer speeds attenuated the problem of the high computational cost of Monte Carlo simulations.

In the following subsections, a detailed description for building a solution for the RTE, step by step, will be shown.

As seen in [10], there are four main Monte Carlo methods for photon migration in turbid media, that is, four different ways to build photon's trajectories. The four methods are referred as the albedo-weight method (AW), the albedo-rejection method (AR), the absorption-scattering path length rejection method (ASPR), and the microscopic Beer-Lambert law method (mBLL).

The AW method considers a source emitting packets of many photons and after each interaction a fraction of the photons in the packet are absorbed, i.e., lost. In this type of simulation, the packet is emitted with an initial weight of 1 and at each interaction the current weight is multiplied by the albedo to account for the loss of photons by absorption and the packet continues propagating in the scattered direction with a reduced weight. This method has in its favor that there are no wasted computations because all photon packets eventually reach the detector. This approach was favored by several research papers [11–16].

In this chapter, we present a simulation using the AR method; the main advantage of this method is that it mimics what happens in nature by tracking individual photons emitted by some source. Naturally for this case, the computational time of the absorbed photons are wasted.

It is seen in the investigation conducted by Sassaroli that the total probability of detecting a photon is equivalent in the four methods and they have statistical equivalence, with some differences regarding the convergence time. When comparing the AW to the AR method, as one may expect, the AR will require, in most scenarios, more launched photons for convergence. However, it is also noted that in the AR method, the photon is detected or lost similarly to what happens in a time-resolved experiment in which photons are detected using the time-correlated single photon counting technique and the noise on the simulated response reproduces the Poisson statistics that typically characterize the noise in experiments. Therefore, as we already stated, the AR method reproduces a more physical simulation of

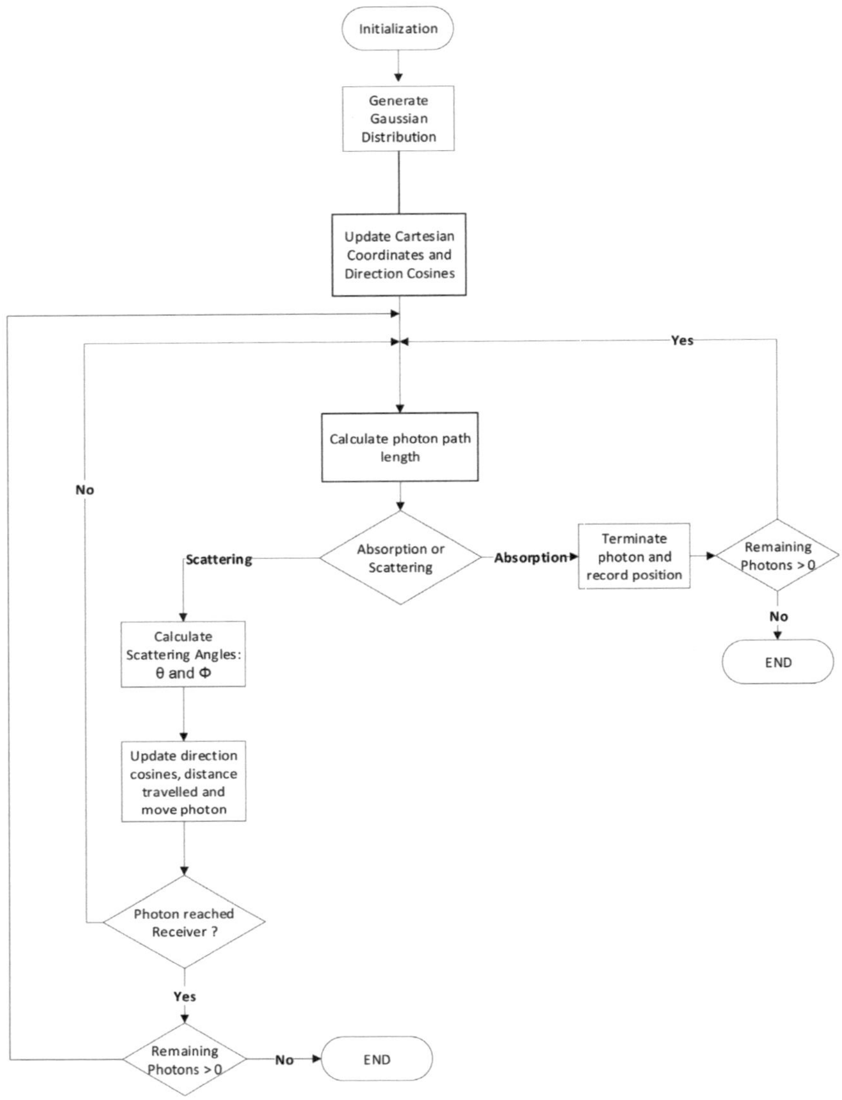

Figure 2.
Flow chart for the AR Monte Carlo simulation.

propagation than the AW method, where the simulated photons are energy packets rather than physical photons. In **Figure 2**, we show a flow chart of the Monte Carlo algorithm used in this simulation, each step in this chart will be explained with further detail in the following sections.

3.1 Photon path length

The photon path length, or step size, determines how far the photon will travel before encountering a water molecule or particle, meaning the distance a photon travels between an absorption event and a scattering event. The path length is directly related to the water type and, consequently, to the beam attenuation coefficient introduced in Section 2.1, and shown for several water types in **Table 1**.

As stated in [4], from the derivation of the RTE, the radiance in a particular direction (θ, ϕ) decays due to absorption and scattering according to:

$$\frac{dL(r,\theta,\phi)}{dr} = -\mathrm{c}(r)\mathrm{L}(r,\theta\phi), \tag{31}$$

where $c(r)$ is the beam attenuation coefficient and r is the distance from a given starting point. Integrating Eq. (31), we have:

$$L(r,\theta,\phi) = L(0,\theta,\phi)e^{-\int_0^r c(r')d(r')}, \tag{32}$$

putting it in terms of the photon path length $\tau = \int_0^r c(r')dr'$:

$$L(\tau,\theta,\phi) = L(0,\theta,\phi)e^{-\tau}. \tag{33}$$

The exponential decay of radiance can be explained in terms of the fate of individual photons if the probability of any photon being absorbed or scattered out of the beam between photon path lengths τ and $\tau + d\tau$ is:

$$p_T(\tau)d\tau = e^{-\tau}d\tau. \tag{34}$$

Mobley [3] notes that $p_T(\tau)$ satisfies the normalization condition for a probability density function (PDF):

$$\int_{x_1}^{x_2} p_x(x)dx = 1, \tag{35}$$

with $x_1 = 0$ and $x_2 = \infty$. The corresponding cumulative distribution function (CDF) is:

$$P_T(\tau) = 1 - e^{-\tau}, \tag{36}$$

selecting a random number q from a uniform distribution, $0 \le q \le 1$, and solving for τ, gives:

$$q = P_T(\tau) = 1 - e^{-\tau}, \tag{37}$$

$$\tau = -\ln(1-q) = -\ln(q). \tag{38}$$

For homogenous water, where $c(r)$ does not depend on r, where $\tau = cr$, the photon will travel a geometric distance, r:

$$r = -\frac{1}{c}\ln(q). \tag{39}$$

3.2 Absorption events and scattering angles

Recalling that the photon path length, or step size, is the geometric distance a photon will travel between a scattering event and an absorption event, after determining the distance the photon will travel from origin, one must decide if the next interaction will be an absorption or a scattering event. For this, we recall the definition of the single scattering albedo, ω_o, that is the ratio between the scattering and the absorption coefficients. The single scattering albedo is referred sometimes as the probability of photon survival, because this ratio is the probability that the photon will undergo scattering instead of absorption. As we have mentioned before, in this Monte Carlo simulation, we used the AR method and to numerically implement this, we must select again a random number from a uniform distribution, $0 \le q \le 1$, to determine which event will take place:

$$\text{event} = \begin{cases} \text{absorption, for } q \le \omega_o \\ \text{scattering, for } q > \omega_o \end{cases}. \tag{40}$$

From this description, one may clearly see where the name "Albedo-Rejection" comes from: if the photon is absorbed, it must be terminated, and its position is recorded for further analysis; if the photon was scattered, the scattering angles must be calculated, and its direction of propagation is updated.

3.2.1 The Henyen-Greenstein phase function

Unlike the FSO channel, in an underwater environment, the photons will encounter a much larger number of particles, and thus multiple scattering events will play a more important role in the received optical power, when compared to other optical wireless communication channels.

One of the models proposed to describe the SPF is the Henyen-Greenstein phase function (HG) [12, 17, 18]. The HG was originally proposed by Henyen and Greenstein to describe the scattering of light by interstellar dust [19]:

$$\tilde{\beta}(g,\theta) = p_{HG} = \frac{1}{4\pi}\frac{1-g^2}{(1+g^2-2g cos\theta)^{\frac{3}{2}}}, \tag{41}$$

where g is referred to as the anisotropy factor and is also the mean cosine of the scattering angle $\cos\theta$, which takes values between -1 and 1. When $g = -1$, there is complete backscattering; for $g = 0$, the scattering is isotropic; and $g = 1$ gives complete forward scattering. Using $g = 0.924$, the HG is a good approximation for photon propagation in water and a good fit for the Petzold's measurements.

To find the scattering angle, Eq. (41) must be solved for θ:

$$\cos\theta = \frac{1}{2g}\left\{1+g^2-\left(\frac{1-g^2}{1-g+2gq}\right)^2\right\}, \tag{42}$$

where q is a uniformly distributed random number between 0 and 1. The complete method for obtaining Eq. (42) may be found elsewhere [4].

Because scattering is a three-dimensional process, besides the polar angle (θ) defined above, the azimuthal (ϕ) angle must also be defined. For an unpolarized beam, the azimuthal angle has an equal probability to take any value between 0 and 2π [3]. Selecting again a random number, the azimuthal angle ϕ can be found by:

$$\phi = 2\pi q. \tag{43}$$

Comparing the HG phase function using Eq. (41) with the Petzold's avg. Phase Function extracted from [3], we can see in **Figure 3** that the HG is indeed an adequate approximation for the SPF.

The HG phase function found great use in ocean optics simulations and is heavily used in Monte Carlo simulations and other relevant channel modeling works by several researchers [20].

3.2.2 The two-term Henyey-Greenstein phase function

Despite being an approximate fit, it can be seen in **Figure 4** that the HG phase function shows noticeable differences from the Petzold's measurements for small angles, $\theta < 20^\circ$, and large angles $\theta > 150^\circ$.

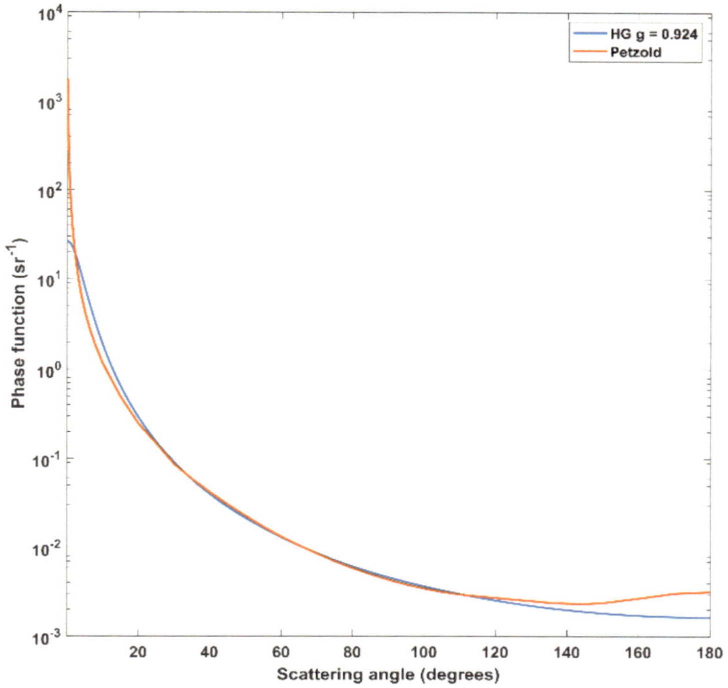

Figure 3.
Comparison of the HG (blue) and the Petzold's average phase function (red).

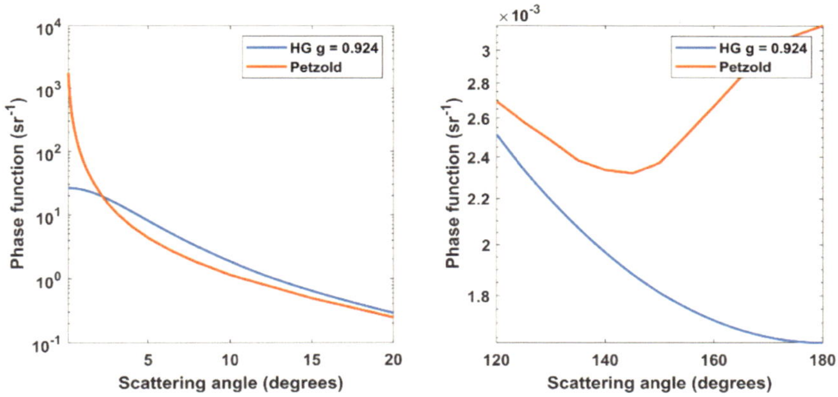

Figure 4.
Comparison between the HG and Petzold's average phase function (left) and large (right) scattering angles.

To overcome this limitation, Haltrin [21] proposed a two-term Heyen-Greenstein (TTHG) phase function that matched Petzold's results better at small scattering angles and provided the elongation seen at larger angles. The TTHG is given by:

$$p_{TTHG}(\theta) = \alpha p_{HG}(\theta, g_1) + (1 - \alpha) p_{HG}(\theta, g_2), \tag{44}$$

where p_{HG} is the HG given by Eq. (41), α is a weighting factor, and g_1 is given a value near 1, which will make the TTHG increase more strongly at small angles. The parameter g_2 takes a negative value that will make the TTHG increase, as the angle θ approaches 180°. The values of α and g_2 may be calculated by:

$$g_2 = -0.3061446 + 1.000568 g_1 - 0.01826332 g_1^2 + 0.03643748 {g_1}^3, \quad (45)$$

$$\alpha = \frac{g_2(1+g_2)}{(g_1+g_2)(1+g_2-g_1)}. \quad (46)$$

Haltrin also derived a relationship that relates the scattering coefficient, b, with the anisotropy factor, g_1:

$$g_1 = 1 - \frac{0.001247}{b}. \quad (47)$$

After calculating the parameters using Eqs. (44)–(47) and computing the phase function for clear ocean waters with $b = 0.037\ m^{-1}$ using Eq. (44), we now compare the resulting phase function with the Petzold's phase function for clear ocean water. We can see from **Figure 5** that the TTHG represents a better approximation for small angles and presents the characteristic elongation at large angles. Despite that, in **Figure 6**, it can also be seen that the TTHG with $g_1 = 0.9943$ calculated using Eq. (47), differs greatly from the Petzold's average phase function data at intermediate angles, 20–110°. Comparing the results for the TTHG for different

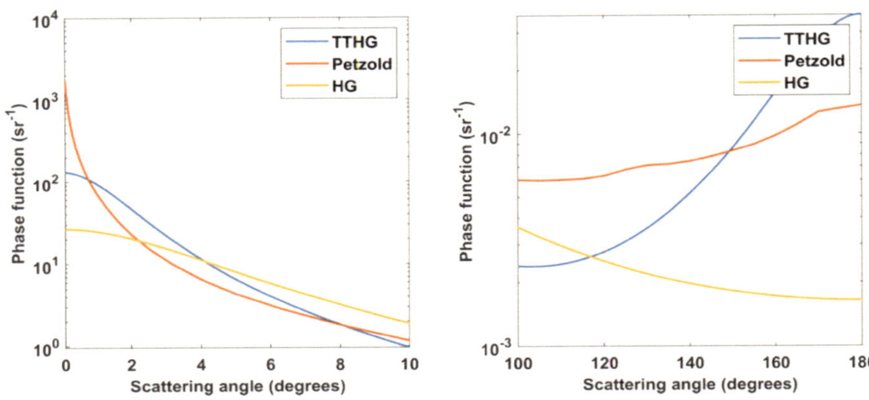

Figure 5.
Comparison of small and large angles for two phase functions, TTHG and HG, with Petzold's data.

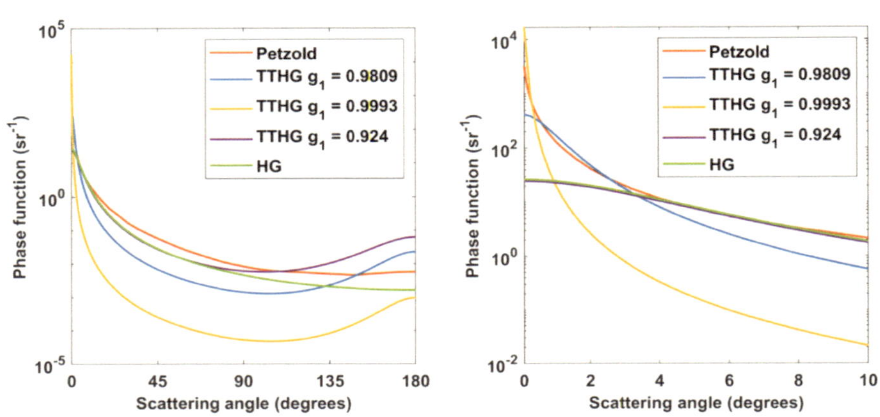

Figure 6.
Comparison of scattering angles for two phase functions with Petzold's data for different values of the anisotropy factor g_1.

values of g_1 against the Petzold's average phase function, we can see that $g_1 = 0.9809$, as proposed in [17], provides a good fit at small angles, where the phase function has its largest values, and maintains the elongation at larger angles characteristic of the Petzold's phase function.

3.2.3 Haltrin empirical phase function

Despite the attempts from several researchers of developing analytical phase functions for natural waters, we can see that the ones described in the previous sections do not provide a good fit over the total range of θ. In [22], Kirk computed a CDF from Petzold's data for θ over the range of 2.5–180° in intervals of 5°. Unfortunately, his CDF did not include data for angles smaller than 2.5°, angles where, as seen before, a lot of scattering happens. In the results presented later in Section 4, we computed a CDF of Petzold's data over the whole range of scattering angles and used linear interpolation to compute θ for each random value selected. The accuracy obtained by this method comes with the drawback of increased computation time.

In [23], Haltrin developed an empirical fit using strong regression for all Petzold's measurements. One of the strongest points of this model is that the Phase Function may be obtained only by using the scattering coefficient and the single scattering albedo as inputs:

$$\beta(\theta) = \exp\left[\psi\left(1 + \sum_{n=1}^{5} (-1)^n k_n \theta^{\left(\frac{n}{2}\right)}\right)\right], \tag{48}$$

and the scattering phase function:

$$p(\theta) = \frac{4\pi}{b} \exp\left[\psi\left(1 + \sum_{n=1}^{5} (-1)^n k_n \theta^{\left(\frac{n}{2}\right)}\right)\right], \tag{49}$$

with:

$$\frac{1}{2}\int_0^\pi p(\theta) \sin(\theta) d\theta = 1, \tag{50}$$

where b is the scattering coefficient and the remaining parameters are:

$$\psi = 2.598 + 17.748\sqrt{b} - 16.722b + 5.932b\sqrt{b}, \tag{51}$$

$$k_1 = 1.188 - 0.688\omega_0, \tag{52}$$

$$k_2 = 0.1(3.07 - 1.90\omega_0), \tag{53}$$

$$k_3 = 0.01(4.58 - 3.02\omega_0), \tag{54}$$

$$k_4 = 0.001(3.24 - 2.25\omega_0), \tag{55}$$

$$k_5 = 0.0001(0.84 - 0.61\omega_0). \tag{56}$$

The strong regression given by these equations can be used for an empirical model for the phase functions. As seen from **Figure 7**, this model represents a better fit to the Petzold's measurements than the phase functions presented before, especially in the regions where the phase function is the strongest.

In our simulations, we found that when comparing the CDF computed by directly interpolating the Petzold's measurements, the CDF computed from the empirical phase function showed some divergence. Our CDF showed that for

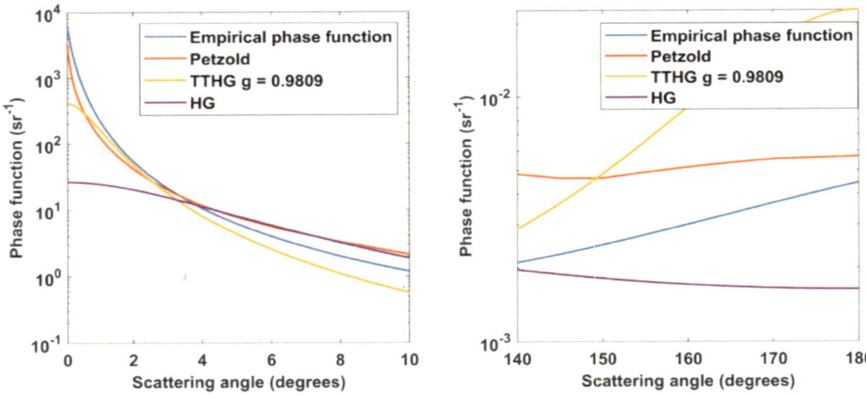

Figure 7.
Comparison of the empirical phase function, the TTHG and HG, against the Petzold's measurements for small and large angles.

harbor waters with $b = 1.8241$ and $\omega_0 = 0.83$, $P(\theta \leq 10^\circ) = 0.64$; while in the empirical phase function, we found $P(\theta \leq 10^\circ) = 0.84$.

The above presented divergence resulted in a small difference in the final received power; however, the empirical phase function largely overestimated the channel bandwidth, especially in harbor waters. This is due to the smaller scattering angles generated by this empirical phase function, causing the photons to travel in a straighter line when compared to the case of scattering with the CDF computed from Petzold's data.

3.2.4 The Fournier-Forand phase function

As seen the previous sections, several analytical and empirical formulae were used as empirical fits for ocean waters but none, specially the analytical ones provide a perfect fit for all scattering angles.

In [24], Fournier and Forand derived an analytic expression for the phase function of a set of particles that have a hyperbolic particle size distribution, with each particle scattering according to the anomalous diffraction theory. The Fournier-Forand (FF) phase function is given by:

$$\begin{aligned}\widetilde{\beta}_{FF}(\theta) = & \frac{1}{4\pi(1-\delta)^2\delta^v}\left\{[v(1-\delta)-(1-\delta^v)]+[\delta(1-\delta^v)-v(1-\delta)]\sin^{-2}\left(\frac{\theta}{2}\right)\right\} \\ & +\frac{1-\delta_{180}^v}{16\pi(\delta_{180}-1)\delta_{180}^v}\left(3\cos^2\theta-1\right),\end{aligned} \tag{57}$$

where

$$v = \frac{3-\mu}{2}, \tag{58}$$

$$\delta = \frac{4}{3(n-1)^2} sin^2\left(\frac{\theta}{2}\right). \tag{59}$$

Being n in Eq. (59) the real index of refraction of the particles, μ in Eq. (58) the slope parameter of the hyperbolic distribution, and δ_{180} in Eq. (57) is δ evaluated at $\theta = 180^\circ$.

It can be seen in [25] that with $n = 1.16$ and $\mu = 3.4586$, the FF phase function gives a good fit to the average Petzold phase function over all scattering angles. This comparison is shown in **Figure 8**.

When comparing the CDF computed from the FF phase function and the CDF from Petzold's data for harbor waters, we get very close probabilities over all scattering angles.

3.2.5 Sahu and Shanmugam (SS) phase function with flat back approximation

In [25], a semi-analytical model for a phase function was proposed by Sahu and Shanmugam (SS). The SS phase function predicts the light scattering properties for all forward angles, namely $0^\circ \leq \theta \leq 90^\circ$ as function of the real index of refraction and the slope parameter of the hyperbolic distribution. The phase function was divided in two parts, one from 0.1 to 5° and the other from 5 to 90°. The authors state that this division was necessary, since the phase function is highly nonlinear with a change of slope around 5°. This phase function is given by:

$$\log\left(\widetilde{\beta}(\theta)\right) = \begin{cases} P_1(\ln(\theta))^2 + P_2(\ln(\theta)) + P_3 \; 0.1^\circ \leq \theta \leq 5^\circ \\ P_1(\ln(\theta))^3 + P_2(\ln(\theta))^2 + P_3(\ln(\theta)) + P_4 \; 5^\circ \leq \theta \leq 90^\circ \end{cases}, \tag{60}$$

where

$$P_m = a_m e^{-x} + b_m(x) + c_m, \tag{61}$$

$$a_m = \frac{d_m}{y^2} + e_m \sin(4\pi y) + f_m y + g_m, \tag{62}$$

$$b_m = \frac{h_m}{y^2} + i_m \sin(4\pi y) + j_m y + k_m, \tag{63}$$

$$c_m = \frac{l_m}{y^2} + o_m \sin(4\pi y) + p_m y + q_m, \tag{64}$$

$$x = \mu - 3, \tag{65}$$

$$y = n - 1, \tag{66}$$

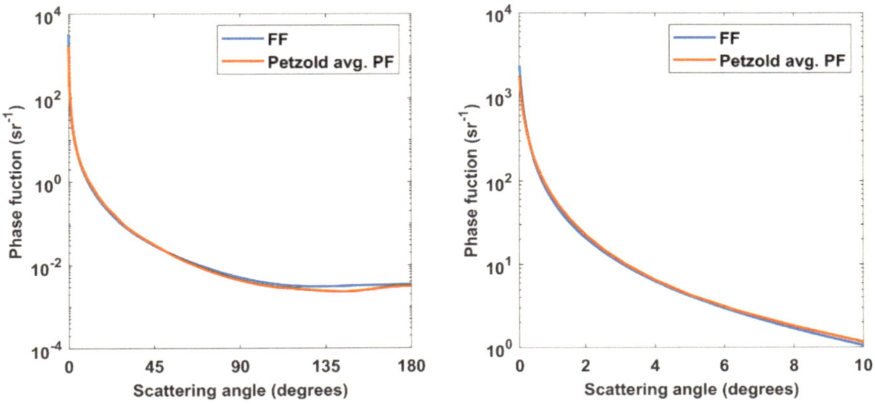

Figure 8.
Comparison of the FF and the Petzold average phase function over all scattering angles and for small scattering angles.

with $m = 1, 2, 3$ for $0.1^{\circ} \leq \theta \leq 5^{\circ}$ and $m = 1, 2, 3, 4$ for $5^{\circ} \leq \theta \leq 90^{\circ}$. The values for the regression coefficients $d_m, e_m, f_m, g_m, h_m, i_m, j_m, k_m, l_m, p_m$, and q_m may be found in [26] and are not shown here for brevity.

Sahu found, in his most recent works [14, 26], that the set of values, $n = 1.16$ and $\mu = 3.4319$ gives the best fit to Petzold average phase function. Using Eqs. (60)–(66) and with the values mentioned above for n and μ, we compared the obtained values with the Petzold average phase function and found that it indeed provides an excellent fit at all scattering angles, but specially at small angles where most of the scattering will happen. Results for both parts of the phase function, 0.1–5° and 5–90°, are presented in **Figure 9**.

3.2.6 Computing phase functions for Monte Carlo simulations

As seen in [27], phase functions satisfy the normalization condition:

$$2\pi \int_0^{\pi} \widetilde{\beta}(\theta) \sin\theta d\theta = 1, \tag{67}$$

and to determine the polar scattering angle:

$$q = 2\pi \int_0^{\theta} \widetilde{\beta}(\theta) \sin\theta d\theta, \tag{68}$$

where q is a uniformly distributed random number between 0 and 1.

However, given the shape of most phase functions, it is more efficient [27] to use the equation for the phase function to build up a table of θ vs. $\widetilde{\beta}(\theta)$, compute a CDF making use of the normalization condition, and then interpolate into the CDF the value of θ for each random number, q, drawn. As seen in Section 3.2.3, Kirk [22] computed a CDF for Petzold's measurements in San Diego Harbor for angles between 2.5 and 177.5° in intervals of 5°.

For the results presented in **Table 2**, we computed a CDF directly from Petzold average phase function extracted from [3] and used the same method described in [27] to obtain values for all angles between 0 and 90°. We chose to proceed with numerical integration for this range, and not from 0 to 180°, because as mentioned in Section 3.2.4, the SS phase function provides values only for forward scattering angles.

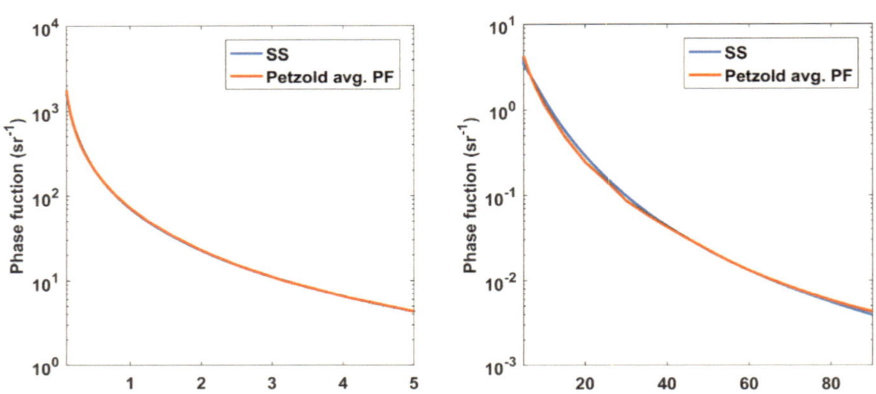

Figure 9.
Comparison between the SS and the Petzold average phase function over the forward scattering angles.

θ (°)	$P(\theta_{PETZ})$	$P(\theta_{FF})$	$P(\theta_{SS})$	$P(\theta_{TTHG})$
0.126	0.009094	0.012355	0.004059	0.002382
0.2	0.031941	0.041314	0.025992	0.011982
0.251	0.046044	0.057930	0.039860	0.020878
0.316	0.062621	0.076605	0.056320	0.034634
0.398	0.081934	0.097344	0.075422	0.055321
0.501	0.104274	0.120254	0.097285	0.085451
0.631	0.129946	0.145647	0.122155	0.127856
0.794	0.158838	0.173561	0.149991	0.184165
1	0.191217	0.204415	0.181067	0.254854
2.512	0.352290	0.358677	0.335353	0.596932
5.012	0.498215	0.505639	0.472517	0.788187
10	0.659192	0.665537	0.630018	0.897428
30	0.891501	0.882151	0.892093	0.974829
50	0.958125	0.951993	0.960280	0.989702
70	0.986110	0.983167	0.987188	0.996168
80	0.994130	0.992771	0.994681	0.998674
85	0.997280	0.996631	0.997562	0.999169
90	1	1	1	1

Table 2.
Computed CDF for selected phase functions.

In **Table 2** and **Figures 10** and **11**, we compare the CDF obtained for three phase functions, the SS, FF, and the TTHG with $g = 0.9809$ against the CDF for Petzold's average phase function. In **Figure 11**, we show the abscissa in logarithmic scale for a better visualization of the differences between phase functions for small values of θ.

In this table, we can see that both the FF and SS phase functions are a great approximation for the Petzold's average phase function, while the TTHG heavily overestimates the amount of scattering up to 10° and underestimates the amount of scattering in intermediate angles, i.e., between 30° and 80°. As one may expect, the effects of those deviations will be larger for waters where scattering is the main attenuation factor.

3.3 Temporal dispersion and the underwater channel bandwidth

3.3.1 Temporal dispersion

One of the most useful information regarding the underwater channel that may be provided by the Monte Carlo simulation, which is unavailable in most simple models like the Beer law, is the channel temporal dispersion. As seen before in this chapter, seawater is a dispersive medium in which light will suffer the effect of multiple scattering events. Scattering will cause the photons to arrive at the receiver at different time intervals causing temporal dispersion and a penalty in the channel bandwidth. In recent years, several researchers employed the double gamma functions to model the impulse response in underwater channels [15, 28]. Despite the accuracy of the double gamma function models, we find that the method described

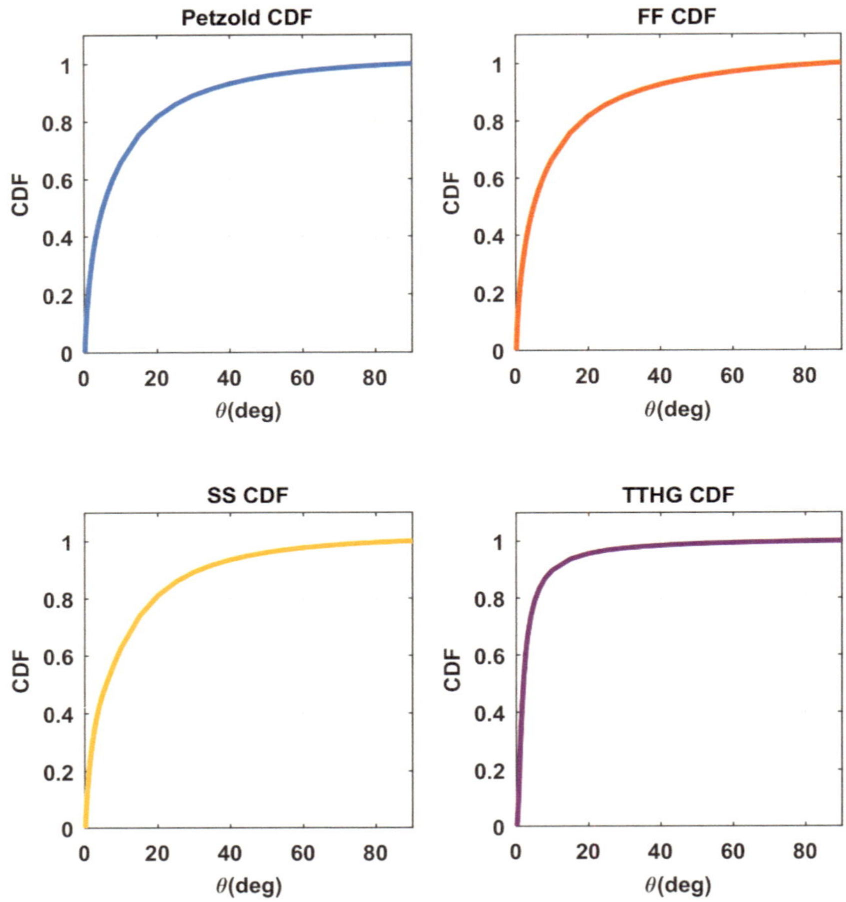

Figure 10.
CDF comparison for selected phase functions.

below consists in a simpler method by means of tracking the individual time of propagation of each photon.

By tracking the propagation distance of each photon (d_{prop}), the time of propagation (TOP), may be estimated by:

$$t_{op} = \frac{d_{prop}}{c_o}, \tag{69}$$

where c_o is the speed of light in the water given by:

$$c_o = \frac{c_{vacuum}}{n_{water}}, \tag{70}$$

with the index of refraction of sea water at 20°C and $\lambda = 500$ nm being $n_{water} = 1.34295$ [3].

For a better visualization of the impulse response, the time delay may be normalized by the straight-line distance, or direct distance, from source to the receiver using:

$$t_{delay} = t_{op} - t_{direct}. \tag{71}$$

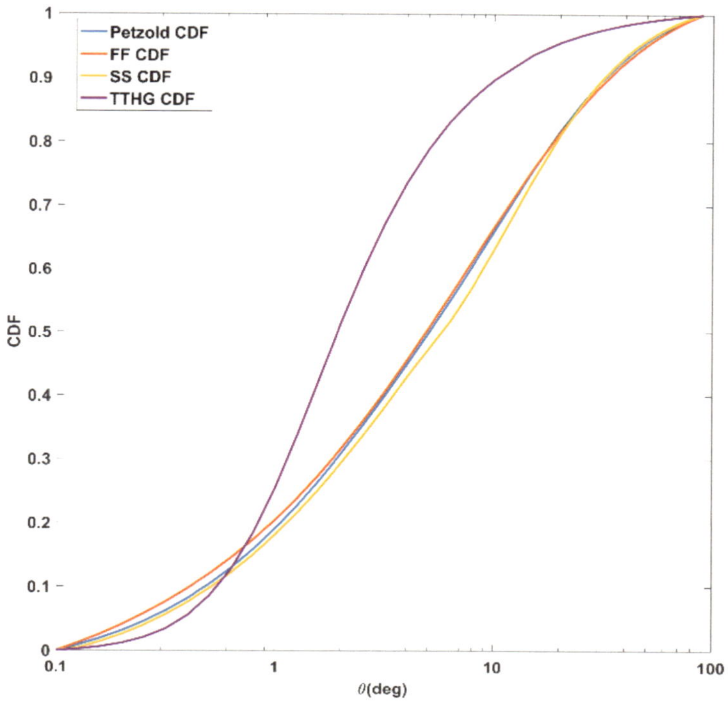

Figure 11.
Log-linear plot of selected phase functions.

Making a histogram of the received photons and distributing the received intensity in time slots defined as the bin widths, one can get a result for the channel impulse response (CIR) of the underwater channel. The results will be presented in Section 4.

3.3.2 Underwater channel transfer function

When using laser diodes in UOWC systems, it is desirable to modulate the light source at very high frequencies. The CIR generated by the Monte Carlo simulation can be converted to frequency domain by the means of a Fourier Transform algorithm, from which we can obtain the channel transfer function.

Given that the impulse response is approximated by a discrete histogram of the time of arrival, it can be transformed to a frequency response, as proposed in [18, 19] by means of a discrete Fourier Transform (DFT), with the usual definition:

$$\mathrm{F}[n] = \mathrm{DFT}[CIR] = \sum_{k=0}^{N-1} t_{\mathrm{delay}}[k] e^{-j\frac{2\pi}{N}nk}, \tag{72}$$

To numerically implement this, the fast Fourier transform (FFT) algorithm can be used to convert the t_{delay} data into the frequency domain:

$$F(f) = \mathrm{abs}\left(fft\left(t_{\mathrm{delay}}(t)dt\right)\right), \tag{73}$$

where dt is the time step, or bin width, used when making the histogram of the time delay data.

As expected, due to Nyquist sampling theorem, the maximum frequency that can be approximated in the frequency domain will be related to the time step by:

$$f_{max} = \frac{1}{2t_{step}}. \tag{74}$$

4. Simulation

In this section, we will provide some simulation results for several emitter parameters, underwater conditions, and different phase functions.

In the first part, we will simulate the received optical power for clear and coastal waters for different beam divergence values and compare it with the popular Beer-Lambert law. It can be seen in [14] that in these water conditions, the delay spread is so small that they can be considered as a nondispersive medium even for high data rate communications, and for this reason the channel impulse response (CIR) will be studied only for conditions where the medium is dispersive.

In the second part, we will study the effect of different phase functions in Harbor waters with a perfectly collimated beam. We believe that this is an interesting approach since in these waters', due to the higher scattering coefficient b, scattering will be a much larger factor and the amount of scattering that occurs at a given angle is an important factor for the received optical power and the delay spread. We will also provide numerical results for the CIR and an estimative for the 3-dB underwater channel bandwidth. The simulation parameters for Sections 4.1 and 4.2 are listed in **Tables 3** and **4**, respectively.

Parameter	**Value**
Simulated beam	Gaussian beam
Beam divergence at $1/e^2$	0, 10, 20 and 30 mrad
Beam width	1 mm
Receiver diameter	0.1 m
Receiver FOV	180°
Photons per simulation	100×10^6
Phase function	Fournier-Forand

Table 3.
Simulation parameters for Section 4.1.

Parameter	**Value**
Simulated beam	Gaussian beam
Beam divergence at $1/e^2$	0
Beam width	1 mm
Receiver diameter	0.1 m
Receiver FOV	180°
Photons per simulation	100×10^6
Phase functions	Petzold average, FF, SS and HG $g = 0.924$

Table 4.
Simulation parameters for Section 4.2.

4.1 Clear and coastal waters

4.1.1 Simulation results

In **Figure 12**, we compare the received power for clear ocean waters with varying beam divergence parameters of 0, 10, 20, and 30 mrad for an emitted beam with a FWHM width of 1 mm. For the phase function, we used the FF, since, as seen in Section 3.2.4, it provides a good fit for Petzold's phase function. However, as mentioned in the "Introduction," when discussing simulation, the impact of different phase functions, especially those which are a good approximation for the Petzold's average phase function, is small for the clear and coastal waters due to the low contribution of scattered photons on the final received power.

Our simulation matches very well the Beer-Lambert law for the case of an emitted Gaussian beam with no divergence, since one of the main assumptions of the law is that the emitted beam is perfectly collimated, and as the beam divergence increases the power loss becomes more accentuated.

We note that for a beam with no divergence, after 10 m, the normalized received optical power would be 2.2×10^{-1} for a receiver aperture of 10 cm. Since the beam spread after a distance z can be calculated by:

$$w(z) = 2z \times \tan\left(\frac{\phi}{2}\right), \tag{75}$$

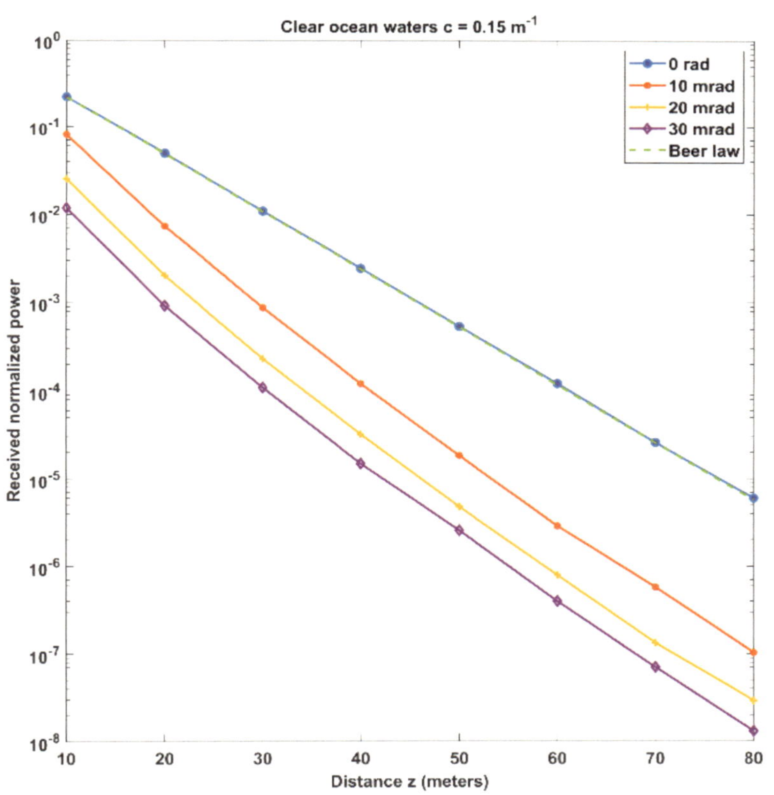

Figure 12.
Received power for clear ocean waters with varying beam divergence parameter.

we may expect a beam spread of 20 cm after 10 m for a beam divergence of 10 mrad, potentially reducing by more than half the received optical power (due to the limited aperture of 10 cm). In fact, we can verify from **Figure 12** that the received power for a 10 mrad divergence is 8.1×10^{-2}, ~40% of the power when the beam was perfectly collimated. The power loss due to the beam spread can be further confirmed by analyzing the position of the received photons along the receiver axis of both configurations, the result being shown in **Figure 13**. From this figure, we can see that for the case of a beam with a divergence of 10 mrad, the photons are spread all over the receiver; while, in the case of a collimated beam, the photons are concentrated along the initial FWHM width of 1 mm.

To further illustrate this, we show, in **Figure 14**, the geometrical losses corresponding to beam divergence. Given that for a perfectly collimated beam in clear ocean waters, power loss is mainly due to absorption and the additional power loss solely due to the spread of the Gaussian beam.

For the case of coastal waters with $c = 0.4m^{-1}$, we can see from **Figure 15** that the Beer law is still a good approximation for a collimated beam, but the law starts to underestimate the received optical power due to the increasing contribution of scattered photons to the final received power.

As was the case for clear waters, the received power drops considerably with an increase in the beam divergence parameter.

Here, the contribution of scattered photons to the final received power increases: after 20 m, 25% of the received power came from scattered photons, and after 30 m, this contribution is elevated to almost 30%. This is further illustrated in **Figure 16**, where the scattering histograms for $z = 20$ and $30\ m$ are depicted, and in **Figure 17**, where we compare the received beam for the same distances in clear and coastal waters. For coastal waters, the effect of scattering is already noticeable at the distance of $z = 10\ m$ and becomes more accentuated at $z = 20\ m$.

Table 5 gives the received normalized power for clear ocean waters for various distances and beam divergence parameters. In the following table (**Table 6**), the same results are given for coastal waters.

4.2 Harbor waters

In this section, we compare the performance of different phase functions for Harbor I and Harbor II waters. First, we interpolated values for the Petzold average

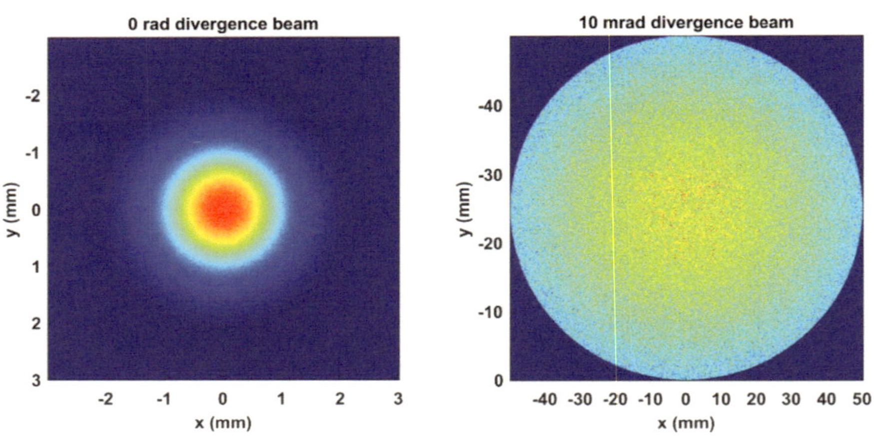

Figure 13.
Beam spread in clear waters for a receiver located at z = 10 m. 0 rad (left) and 10 mrad (right).

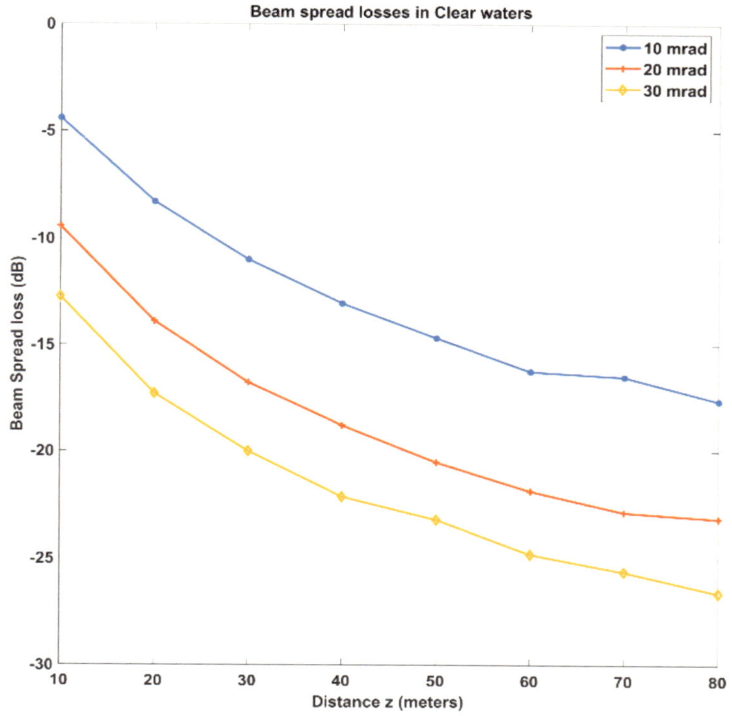

Figure 14.
Beam spread loss in clear waters for beam divergence parameters of 10, 20, and 30 mrad.

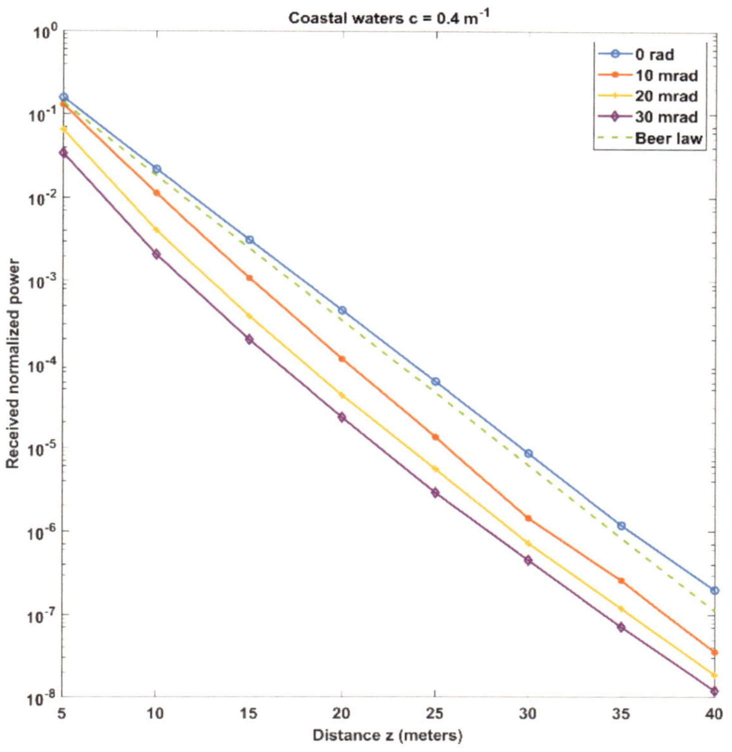

Figure 15.
Received power for coastal waters with varying beam divergence parameter.

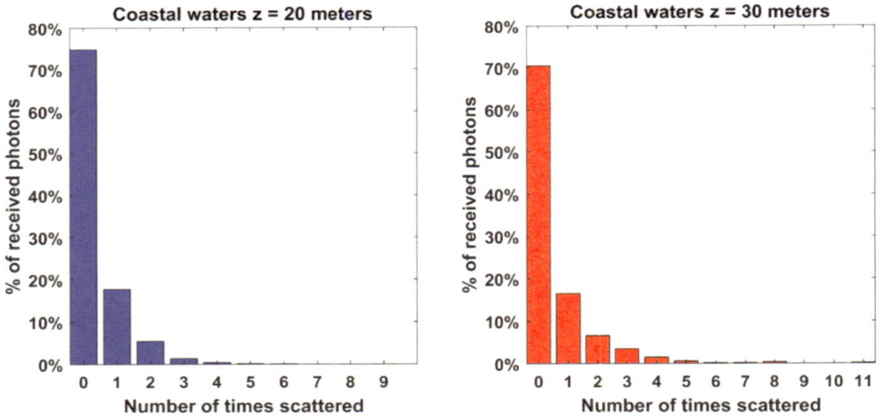

Figure 16.
Scattering histogram for coastal waters with a perfectly collimated beam.

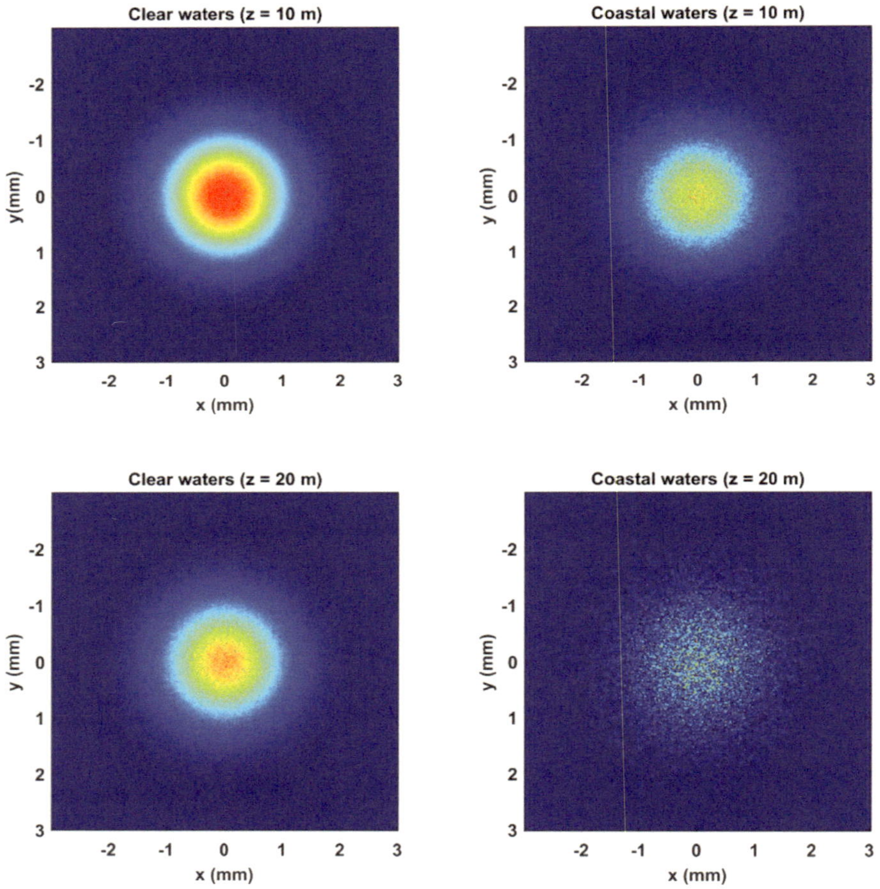

Figure 17.
Received beams in clear and coastal waters.

phase function and considered, as in Section 3.2.5, the benchmark for all other phase functions. The amount of unique scattering angles generated by this interpolation was enough for each photon to have a unique scattering angle.

	Distance			
ϕ **(mrad)**	**10 m**	**30 m**	**50 m**	**70 m**
0	2.2×10^{-1}	1.09×10^{-2}	5.3×10^{-4}	2.6×10^{-5}
10	8.1×10^{-2}	8.7×10^{-4}	1.8×10^{-5}	5.8×10^{-7}
20	2.5×10^{-2}	2.3×10^{-4}	4.7×10^{-6}	1.3×10^{-7}
30	1.2×10^{-2}	1.1×10^{-4}	2.5×10^{-6}	7×10^{-8}

Table 5.
Received power for clear ocean waters.

	Distance			
ϕ **(mrad)**	**10 m**	**20 m**	**30 m**	**40 m**
0	2.2×10^{-2}	4.4×10^{-4}	8.6×10^{-6}	1.1×10^{-7}
10	1.13×10^{-2}	1.15×10^{-4}	1.4×10^{-6}	3.6×10^{-8}
20	4.1×10^{-3}	4.2×10^{-5}	7.2×10^{-7}	1.9×10^{-8}
30	2.1×10^{-3}	2.3×10^{-5}	4.5×10^{-7}	1.2×10^{-8}

Table 6.
Received power for coastal waters.

The SS phase function, seen in Section 3.2.4, predicts scattering for all forward angles. For the angles between 90 and 180°, we first calculated the backscattering ratio, as given in [25]:

$$B_p = P_1\left(\frac{1}{(n-1)^2}\right)^{P_2}, \tag{76}$$

where

$$P_m = a_m(\mu - 3)^2 + b_m(\mu - 3) + c_m, \tag{77}$$

with the values for a_m, b_m, and c_m being the ones extracted from [25].

Using the values of $n = 1.16$ and $\mu = 3.4319$, the backscattering ratio $B_p = 0.0185$ is obtained, considering the scattering to be uniformly distributed over all angles above 90°, similarly to what was done in [14].

4.2.1 Simulation results

For the received optical power for Harbor I and Harbor II waters, we can see in **Figures 18** and **19** that the results obtained when using the FF and the SS phase functions are closely matched to the results obtained using the interpolated Petzold average PF, meanwhile the one term HG severely underestimates the received optical power. The main reason for this, as we have seen before, is that the HG is unable to reproduce the high amount of scattering that occur at small angles, as is the case in the Petzold average PF and matched by the other analyzed phase functions.

In **Figure 20**, the received beams are compared for Harbor I and Harbor II waters for $z = 4$ and $6\ m$. This figure clearly shows the higher scattering coefficient of Harbor II waters and how it impacts the received beam just after 4 m.

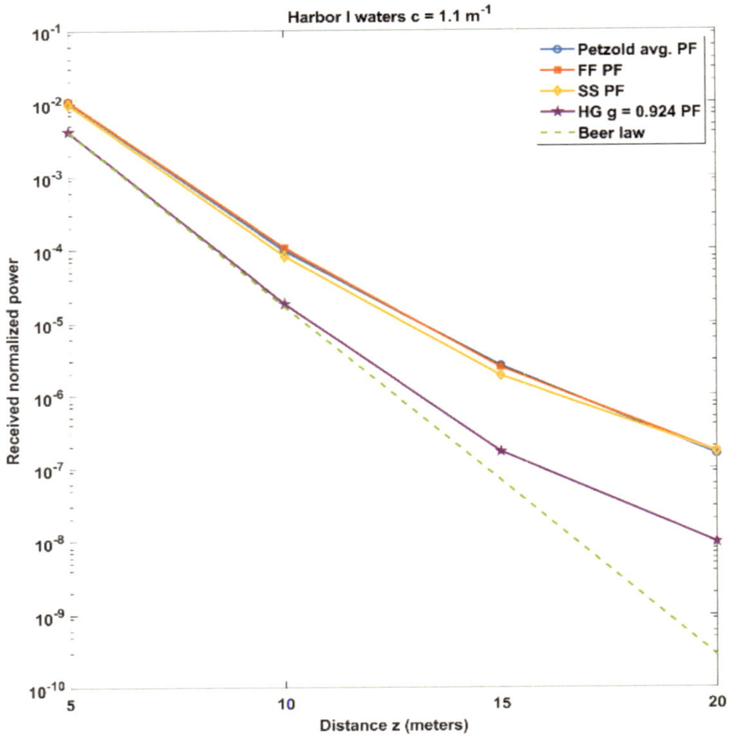

Figure 18.
Received power for Harbor I waters for selected phase functions.

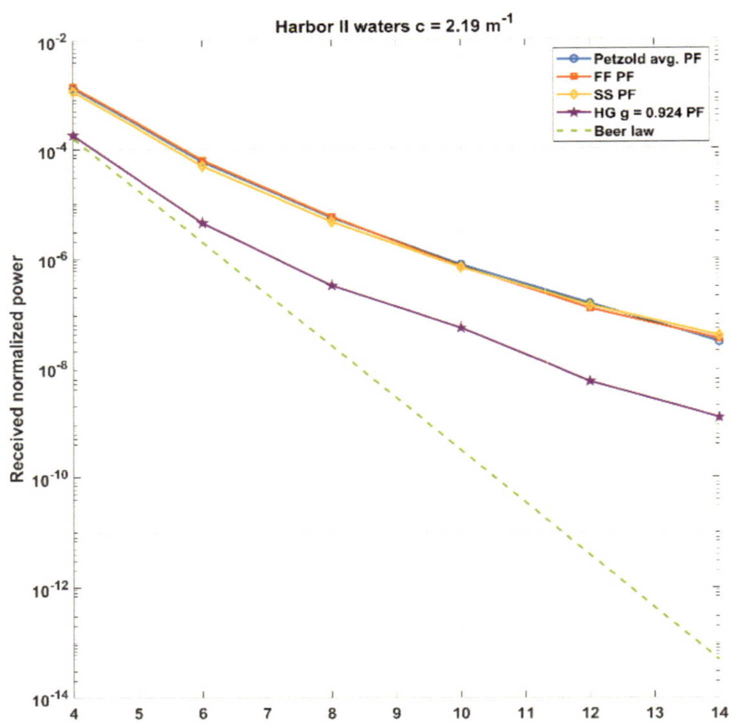

Figure 19.
Receiver power for Harbor II waters for selected phase functions.

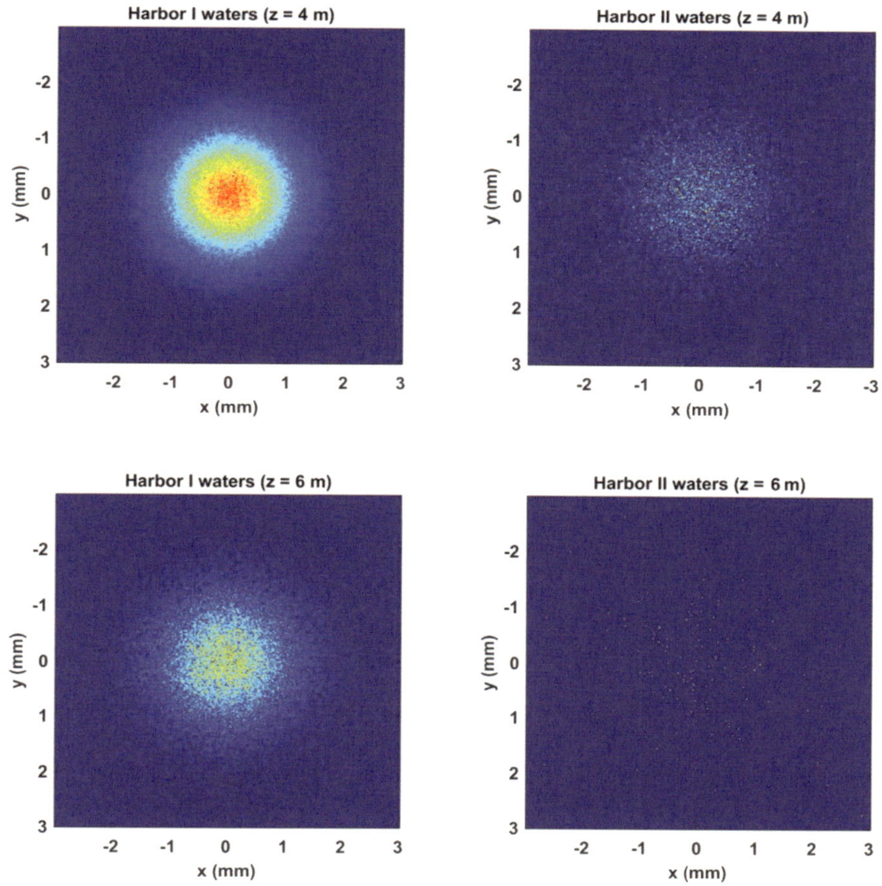

Figure 20.
Received beams in Harbor I and Harbor II waters.

When looking at the CIR for Harbor I waters with z = 15 m, plotted in **Figure 21**, we can see that the results for the Petzold, FF, and the SS phase functions show similar behavior on the delay profile, while the CIR for the HG phase function shows a considerable number of delayed photons.

Figure 22 plots the transfer function of the channel, from which the 3-dB bandwidth can be extracted, for Harbor I waters corresponding to z = 15 and 20 m using the method described in Section 3.3.2. It is possible to verify from these results that the behavior of the SS phase function closely matches the one for the Petzold PF for both distances and the HG underestimates the bandwidth by a large margin in both scenarios. The FF phase function on the other hand, overestimates the bandwidth for the case z = 15 m. The 3-dB bandwidth is limited to $\sim$300 and 90 MHz, for 15 and 20 m, respectively.

The same analysis is now applied to Harbor II waters. The impulse response is shown in **Figure 23**. The interpretation is that the HG phase function exhibits a response where the power is much more distributed over time when comparing to what is seen in the other phase functions where most of the power is located at the instant $t_{direct} = 44\ ns$.

The effect of the delay spread is even more noticeable when we analyze the transfer function. In **Figure 24**, this analysis is done for z = 10 and 12 m, where we see that in these conditions, due to higher number of scattering events that each photon undergoes, the bandwidth is lower than what we found at Harbor I waters, even when considering smaller propagation distances. In this case, the phase

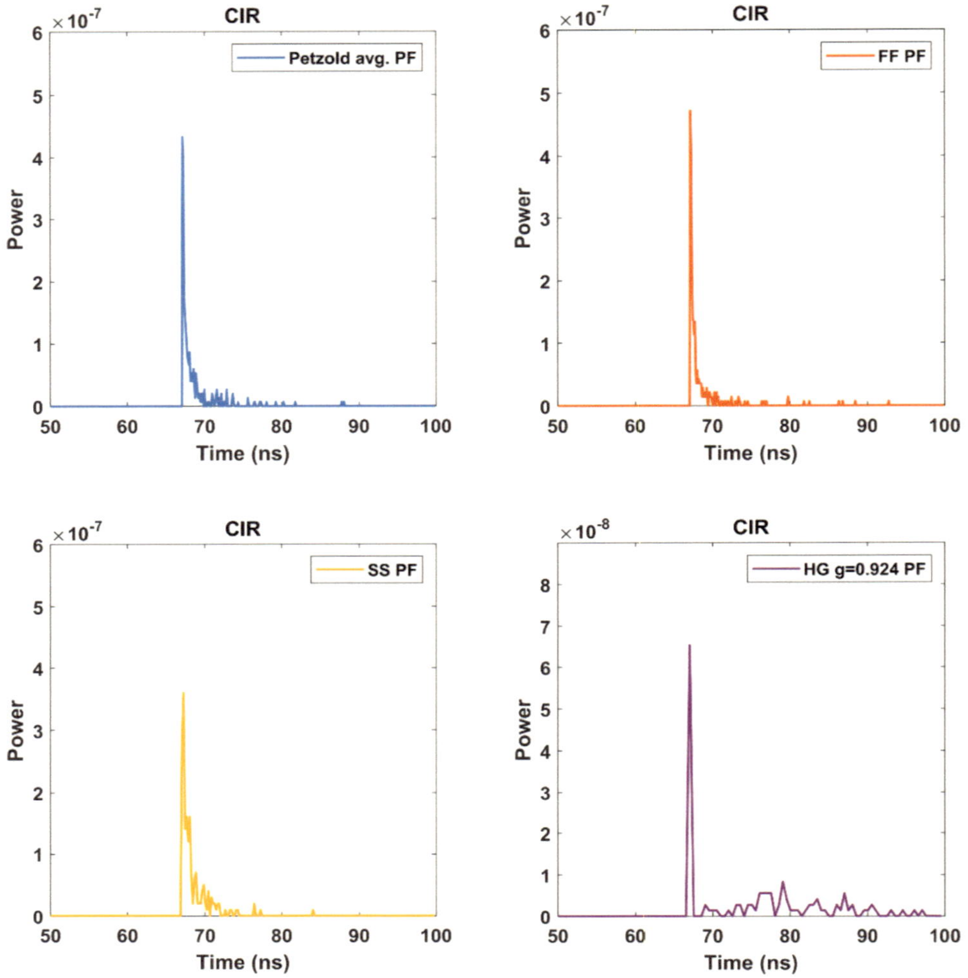

Figure 21.
Channel impulse response for Harbor I waters with z = 15 m.

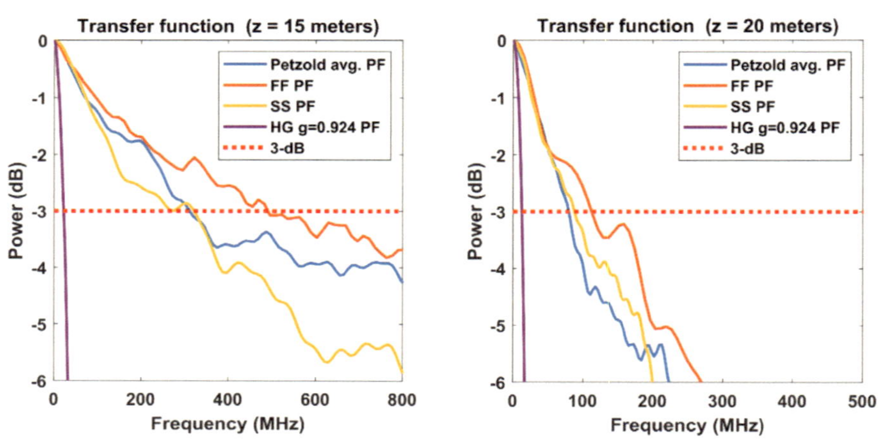

Figure 22.
Transfer function for Harbor I waters. z = 15 and 20 m.

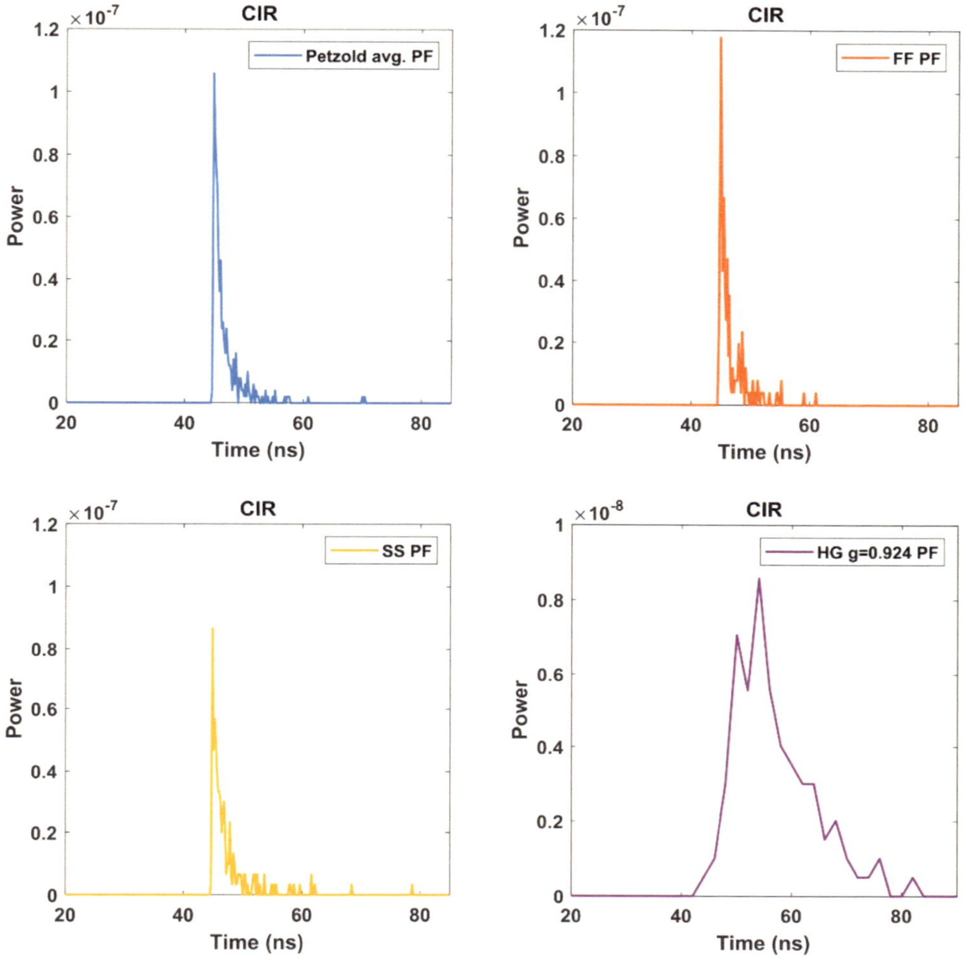

Figure 23.
Channel impulse response for Harbor II waters with z = 10 m.

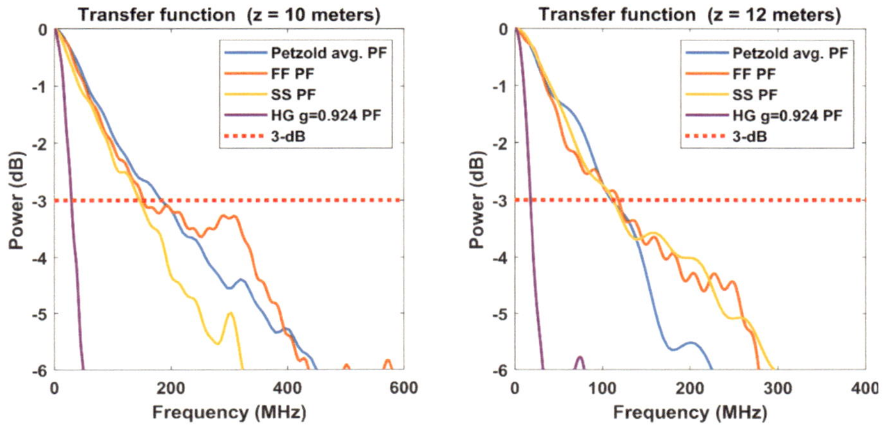

Figure 24.
Transfer function for Harbor II waters. z = 10 and 12 m.

	Received power	τ_{RMS} (ns)	3-dB BW (MHz)
HI z = 15 m	2.6×10^{-6}	2.93	310
HI z = 20 m	1.8×10^{-7}	6.18	80
HII z = 8 m	5.4×10^{-6}	2.06	440
HII z = 10 m	7.7×10^{-7}	3.51	185
HII z = 12 m	1.5×10^{-7}	4.78	110
HII z = 14 m	2.8×10^{-8}	7.87	53

Table 7.
Results for received power, rms delay, and 3-dB BW for Harbor I and Harbor II waters.

functions FF and SS give similar results to the Petzold average PF, the bandwidth being limited to ~150 and 100 MHz, for distances of 10 and 12 m, respectively.

In the following table, **Table 7**, we summarize the most relevant numerical results for Harbor I and Harbor II waters as we did for the case of clear and coastal waters in Section 4.1. For brevity, we only show the results obtained using the Petzold average phase function; however, as noted in the results presented previously, one may expect similar numerical results when using the FF or the SS phase function as they predict scattering angles similar to those predicted by Petzold.

4.3 Validation of simulation results

For further validation of the simulation model, we compared the values obtained in our simulation with the ones obtained by Sahu in [14]. For this, we used the same parameters used in his simulations, a Gaussian beam profile with FWHM beam width of 2 mm, a beam divergence parameter of 1.5 mrad, and a receiver aperture of 10 cm.

The comparison for clear and coastal waters between our simulation and [14] is shown in **Figure 25**, and we can see that our simulation closely matches the results obtained by Sahu.

For Harbor waters, Sahu only considered Harbor II waters with $c = 2.19\ m^{-1}$, hence we could not compare the values we obtained for Harbor I waters with $c = 1.1m^{-1}$. The results are illustrated in **Figure 26**, where a good agreement is seen.

Furthermore, Sahu defined a delay spread quantity which is the time period over which the CIR falls to −20 dB below its peak. He provided numerical values for

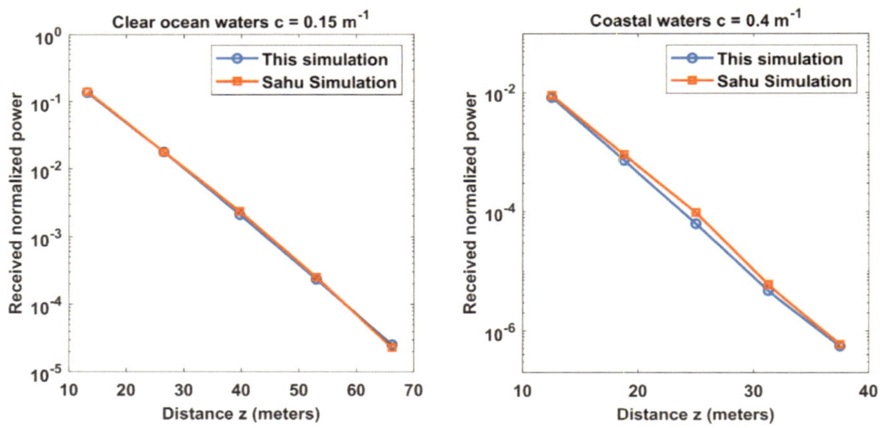

Figure 25.
Comparison between this simulation model and [14].

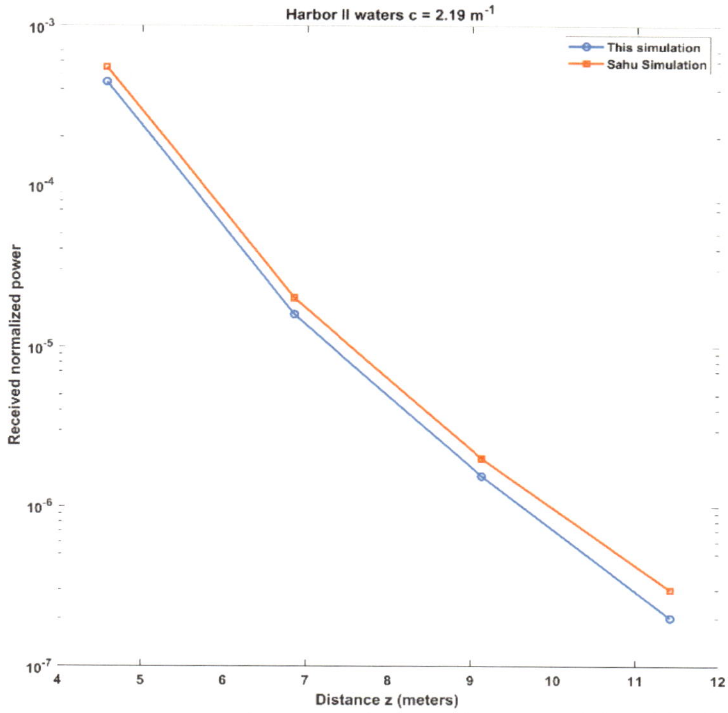

Figure 26.
Comparison between this simulation model and [14].

different propagation distances and the comparison against his intercept points for these distances are depicted in **Figure 27**. The CIR in that figure is normalized by the straight-line time of propagation (t_{direct}) for a better visualization of the delay spread. The black line in the figure is the −20 dB intercept point as given in [14], and we can see that again the values obtained in the simulation are quite close to the ones obtained by Sahu, which validate the model proposed here.

5. Conclusion

In this chapter, a simple yet powerful tool for modeling the propagation of photons in an underwater channel is presented, which provides more accurate results than the conventional Beer-Lambert law. The algorithm presented here highlights the fundamental processes of absorption and scattering, hopefully helping the reader to better understand the physics of photon propagation in an underwater environment.

In the first part of the simulation, we showed how important the collimation of a beam is and how the beam spread can cause losses up to −13 dB even after only 10 m. It was also possible to verify that in clear waters, a good signal to noise ratio is achievable for several tens of meters in both clear and coastal waters.

In the second part of the simulation, the impact of the water turbidity on the underwater link was assessed, where it was concluded that increasing turbidity limits not only the received optical power, but also the maximum bandwidth of the channel. An adequate choice of phase function is important as simpler phase functions, such as the Henyey-Greenstein, may wrongly predict the numerical results in these conditions.

For the first time results for the received power, channel impulse response and 3-dB bandwidth are obtained from Monte Carlo RTE simulations for different phase functions. The SS and FF are compared with the Petzold average PF and results are shown to

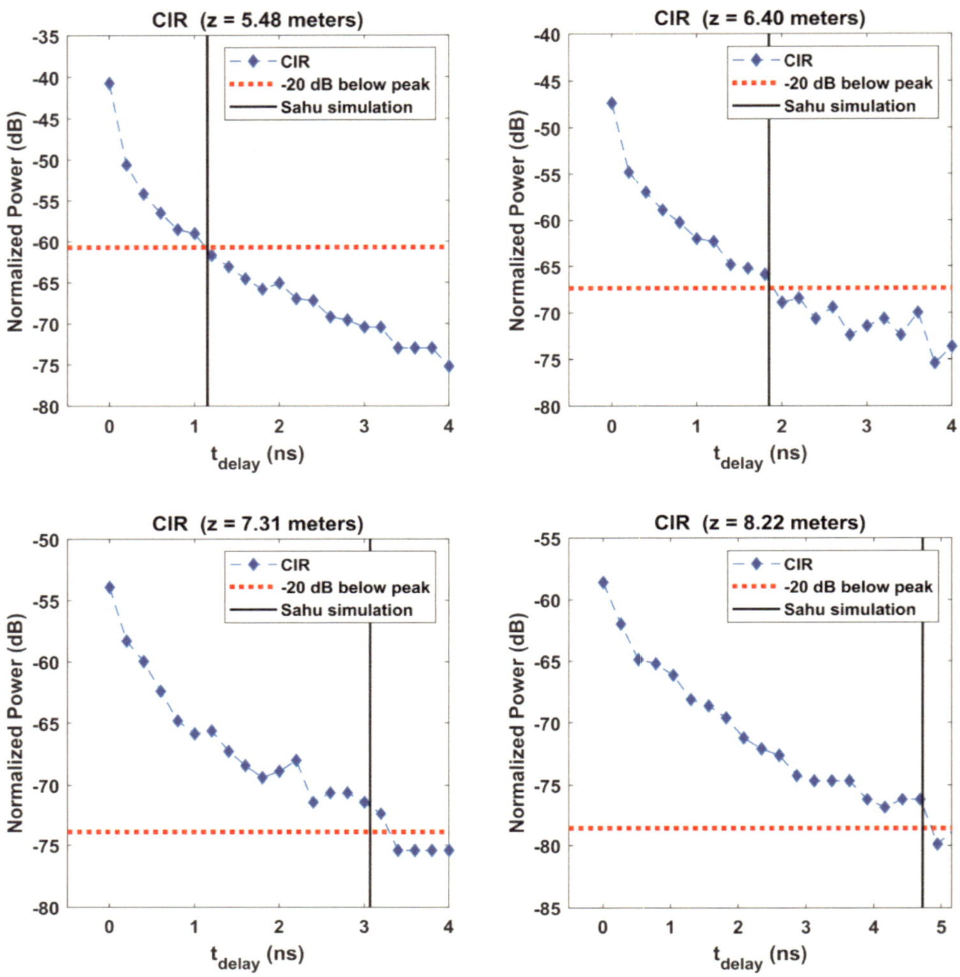

Figure 27.
CIR for various distances in Harbor II waters, where the delay spread is highlighted.

agree well. Moreover, for validation purposes, our simulation results are further compared with those obtained by Sahu [14] and good agreement is shown to exist.

We believe that the approach presented here provides a general and flexible technique for numerically solving the radiative transfer equation, accessible to any reader with basic programming skills. Besides the conditions simulated in this chapter, the simulation program may be used to simulate innumerous other conditions, like misalignment between receiver and transmitter, smaller aperture sizes, limited FOVs, and other light sources other than Gaussian beams.

Acknowledgements

This work is financed by the ERDF—European Regional Development Fund through the Operational Programme for Competitiveness and Internationalisation—COMPETE 2020 Programme, and by National Funds through the FCT—Portuguese Foundation for Science and Technology, I.P., within project POCI-01-0145-FEDER-031971.

Author details

Rafael M.G. Kraemer[1,2], Luís M. Pessoa[2] and Henrique M. Salgado[1,2]*

1 Faculty of Engineering, University of Porto, Portugal

2 INESC TEC—INESC Technology and Science (Formerly INESC Porto), Porto, Portugal

*Address all correspondence to: henrique.salgado@inesctec.pt

References

[1] Inherent Optical Properties. Ocean Optics Web Book [Online]. Available from: http://www.oceanopticsbook.info/view/overview_of_optical_oceanography/inherent_optical_properties [Accessed: 28 September 2018]

[2] Zeng Z, Fu S, Zhang H, Dong Y, Cheng J. A survey of underwater optical wireless communications. IEEE Communication Surveys and Tutorials. 2017;**19**(1):204-238

[3] Mobley CD. Light and Water—Radiative Transfer in Natural Waters. San Diego: Academic Press; 1994

[4] Leathers RA, Downes TV, Davis CO, Mobley CD. Monte Carlo Radiative Transfer Simulations for Ocean Optics: A Practical Guide. Fort Belvoir, VA: Defense Technical Information Center; 2004

[5] The Volume Scattering Function. Ocean Optics Web Book [Online]. Available from: http://www.oceanopticsbook.info/view/overview_of_optical_oceanography/the_volume_scattering_function [Accessed: 30 September 2018]

[6] Petzold TJ. Volume Scattering Functions for Selected Ocean Waters. Fort Belvoir, VA: Defense Technical Information Center; 1972

[7] Metropolis N, Ulam S. The Monte Carlo method. Journal of the American Statistical Association. 1949;**44**(247):335-341

[8] Computing and the Manhattan Project. Atomic Heritage Foundation [Online]. Available from: https://www.atomicheritage.org/history/computing-and-manhattan-project [Accessed: 22 October 2018]

[9] Monte Carlo and the Bomb [Online]. Available from: http://jmoses.co/2014/05/27/monte-carlo-and-the-bomb.html [Accessed: 01 November 2018]

[10] Sassaroli A, Martelli F. Equivalence of four Monte Carlo methods for photon migration in turbid media. Journal of the Optical Society of America A. 2012;**29**(10):2110

[11] Cox WC, Jr, Simulation, Modeling, and Design of Underwater Optical Communication Systems. PhD Thesis, North Carolina State University; 2012

[12] Gabriel C, Khalighi M-A, Bourennane S, Léon P, Rigaud V. Monte-Carlo-based channel characterization for underwater optical communication systems. Journal of Optical Communications and Networking. 2013;**5**(1):1

[13] Li J. Monte Carlo study on pulse response of underwater optical channel. Optical Engineering. 2012;**51**(6):066001

[14] Sahu SK, Shanmugam P. A theoretical study on the impact of particle scattering on the channel characteristics of underwater optical communication system. Optics Communications. 2018;**408**:3-14

[15] Li Y, Leeson MS, Li X. Impulse response modeling for underwater optical wireless channels. Applied Optics. 2018;**57**(17):4815

[16] Vadakke-Chanat S, Shanmugam P, Sundarabalan B. Monte Carlo simulations of the backscattering measurements for associated uncertainty. Optics Express. 2018;**26**(16):21258

[17] Mobley CD, Sundman LK, Boss E. Phase function effects on oceanic light fields. Applied Optics. 2002;**41**(6):1035

[18] Wang L, Jacques SL. MCML—Monte Carlo modeling of light transport in multi-layered tissues in standard C. 183. Computer Methods and Programs in Biomedicine. Boston, MA: Springer; 1995;**47**(2):131-146

[19] Henyey L, Greenstein J. Diffuse radiation in the galaxy. Astophysical Journal. 1941;**93**:70-83

[20] Miramirkhani F, Uysal M. Visible light communication channel modeling for underwater environments with blocking and shadowing. IEEE Access. 2018;**6**:1082-1090

[21] Haltrin VI. One-parameter two-term Henyey-Greenstein phase function for light scattering in seawater. Applied Optics. 2002;**41**(6):1022

[22] Kirk JT. Monte Carlo Procedure for Simulating the Penetration of Light into Natural Waters. Australia: Commonwealth Scientific and Industrial Research Organization; 1981

[23] Haltrin VI. Theoretical and empirical phase functions for Monte Carlo calculations of light scattering in seawater. In: Presented at the Fourth International Conference on Remote Sensing for Marine and Coastal Environments. 1997;**17**:19

[24] Fournier GR, Luc Forand J. Analytic phase function for ocean water. Proc. SPIE Ocean Optics XII. 1994;**2258**:2258

[25] Sahu SK, Shanmugam P. Semi-analytical modeling and parameterization of particulates-in-water phase function for forward angles. Optics Express. 2015;**23**(17):22291

[26] Sahu SK, Shanmugam P. Scattering phase function for particulates-in-water: Modeling and validation. In: Presented at the SPIE Asia-Pacific Remote Sensing; New Delhi, India; 2016. p. 98821H

[27] Mobley CD et al. Comparison of numerical models for computing underwater light fields. Applied Optics. 1993;**32**(36):7484

[28] Tang S, Dong Y, Zhang X. Impulse response modeling for underwater wireless optical communication links. IEEE Transactions on Communications. 2014;**62**(1):226-234

Chapter 9

Energy Aware Router Placements Using Fuzzy Differential Evolution

G. Merlin Sheeba

Abstract

The increasing demand of communication services have led to the increase in energy consumption. Energy sustainability is important and challenging research in current world. An energy aware nearest cell association algorithm is proposed to make the mesh routers (MRs) to sleep if they are in idle state. If the MRs have no associated clients, then the MR is considered to be idle. Any network device in idle state consumes power hence a sleep mechanism is introduced to place energy aware routers. A fuzzy differential evolution (FDE) is introduced to dynamically decide the state of the MR by gaining the knowledge from the fuzzy table for parameters like traffic load, minimum distance and transmission power. Transmission cost and failure rate of the deployed network is evaluated and their performance is analyzed.

Keywords: fuzzy differential evolution, mesh router, mesh client, energy consumption, failure rate, client association, transmission cost

1. Introduction

Future wireless network will make use of more renewable energy sources like solar, wind and hydro. Developing a sustainable communication is one of the critical and challenging issues in order to sustain the ever-growing traffic demands while alleviating the need of increased energy consumption. Traditional solutions have assumed that mesh routers have consistent power supply through wired electricity. Increasing demand for green mesh networks that can harvest their energies via solar or wind power have reached greater attention. But the energy supplies of these rechargeable routers are not consistent but depend on many environmental conditions [1]. For each router, energy consumption is due to the traffic demand of its associated mesh client and the distance between them. The existing router placement algorithms such as, exhaustive search and greedy search [2] are easily running into local optima. To overcome this drawback fuzzy differential evolution (FDE) placement of nodes with Energy Aware Nearest Cell Association (EANCA) algorithm is proposed in this module.

2. System configuration

System model: In this work, any renewable energy powered MRs is considered to be deployed in a grid scenario where the field is divided into grid cells of equal area. The deployed MRs should provide wireless access for all the stationary MCs which are assumed to be distributed using normal probability distribution. The number of MCs associated with the MR changes periodically. Each MR deployed in the candidate location consumes more energy to guarantee QOS and ensure the traffic demands of the associated MCs [3–5].

Energy flow model: The renewable energy powered routers are equipped with battery storage. The continuous time line of energy stored is divided into consecutive slots of 't'. The energy charging and discharging of a router can be defined with a discrete time energy model as [6, 7]

$$E(t) = E(t-1) + H(t) - Ec(t) \tag{1}$$

where $E(t)$ is the residual energy of the router after the tth slot. If $t = 0$, $E(0)$ is the initial stored energy in the router. The harvested energy is denoted as $H(t)$ and the energy consumed is denoted as $Ec(t)$.$H(t)$ is a dynamic function since it purely depends on the nature. Let the maximum charging power of renewable energy powered routers is 100 mW. Here two cases can be analyzed, and the connection failures are adjudged.

$$\text{Case (1) : If } H(t) > Ec(t) \text{ and } E(t) = 1; \text{failure rate is less} \tag{2}$$

$$\text{Case (2) : If } H(t) < Ec(t) \text{and } E(t) = 0; \text{failure rate is more} \tag{3}$$

3. Problem definition

The objective is to place minimum number of energy aware routers such that user's traffic demand is satisfied to minimize energy consumption. Here energy consumption is referred as function of a cost metric called transmission cost (TC). The objective function is given by

$$Minimize\ \ TC = f(D_{min}, TL, P_t) \tag{4}$$

where TC is the transmission cost, D_{min} is the minimum distance between the mesh routers and clients, TL is the traffic load and P_t is transmission power.

Subject to:

$$E_{in} - E_r \geq E_{i,j} \tag{5}$$

where Ein is the initial energy, Er is the residual energy and Ei, j is the energy required to guarantee QOS between the mesh routers and clients.

$$P_{t(i,j)} \leq P_{t(max)} \tag{6}$$

where $P_{t(i,j)}$ is the transmission power required for any ith mesh router to associate with the jth mesh client and $P_{t(max)}$ denotes the maximum transmission power.

$$E_c \leq E_{h+} \tag{7}$$

where E_C is the energy consumed and E_{h+} is the energy harvested.

$$FR \leq F_{\text{th}} \tag{8}$$

where FR indicates the failure rate and Fth denotes the failure threshold.

The FR is expressed as

$$FR = \sum_{t \in K} \sum_{i \in C} \left(1 - \sum_{j \in S} xij(t)\right) \Big/ |C| * |K| \tag{9}$$

where $xij(\mathrm{k}) = \begin{Bmatrix} 1 \; if \; a \; node \; is \; associated \; with \; the \; MR \\ 0 \; otherwise \end{Bmatrix}$, $|C|$ stands for total number of clients, $|K|$ denotes the time slots and S denotes the set of candidate locations [8]. The renewable energy powered routers are equipped with battery storage.

Eq. (5) ensures that the difference between the initial and residual energy must be greater or equal to the energy required to guarantee QOS between the mesh routers and clients. Eq. (6) ensures that the transmission power required for any mesh router to get associated with any client must be less than the maximum transmission power. Eq. (7) ensures that the energy consumed is less than the energy harvested. Eq. (8) ensures that the failure rate (FR) a parameter which is incurred from [8] measures the network performance. The constraint specifies that *FR* must be less than a predefined failure threshold (Fth).

4. Fuzzy differential evolution

According to the former literatures [9] the control parameters S,CR were kept fixed during the optimization process. The control parameters of DE is made adaptive using fuzzy logic. Here in this proposed optimization model fuzzy rules are defined for two sets of input parameters respectively as follows:

1. DE control parameters

2. Network inputs

DE control parameters: In DE algorithm, usually empirical values are selected for its search process. Using fuzzy rules the system is made adaptive to search the parameters for mutation and crossover operation. The setting of parameter values can be done in two ways through parameter tuning and parameter control. Parameter tuning is one of the common methods used to find out good values before running the algorithm. Using the best result the algorithm is made to run. In parameter control approach the values are changed while the algorithm is running [10, 11]. This later approach is further divided into deterministic, adaptive and self-adaptive parameter controls.

Deterministic parameter control: A rule strategy is defined to deterministically change the parameters. It strategically changes the values without getting any feedback from the search.

Adaptive parameter control: This approach is handled when a feedback arises from the search and this feedback is used to determine the values to the parameters.

Self-adaptive parameter control: Here in the evolutionary search encoded parameters are used in the chromosomes to perform the mutation and crossover operation to produce effective individuals that could survive and produce offspring.

Network inputs: In the deployed network the distance between the client and the mesh router in each grid cell is calculated and formulated as a matrix called the distance matrix. The minimum distance value is returned from the matrix which is denoted as Dmin. The Traffic Load (TL) is calculated from the number of clients which are associated with the MRs. The Transmission Power (Pt) needed to associate a MC with MR also varies between low and high.

4.1 Proposed optimization control flow model

Figure 1 shows the fuzzy DE control flow model for optimizing the transmission cost (TC) metric. In general the DE key operators are fixed. In order to overcome the premature convergence which usually occurs in DE, a fuzzy DE approach is implemented wherein the scaling factor (S) and cross over constants (CR) are adapted using the fuzzy inference engine (FIS) [12].

As the network size increases the load of the network, transmission power and the distance between the router and client varies in different instances. The proposed FDE is a knowledge based system which dynamically selects the best search parameters from the fuzzy set. The Mamdani's fuzzy inference method is used to map the output function. The output of each rule is a fuzzy set and the output fuzzy set is the aggregation of all the sets. The sample rule set is constructed and given as follows for the fuzzy inputs CR, S and output F(x) using IF-THEN structure.

IF (CR is low) AND (S is low) THEN F(x) is low.

IF (CR is average) AND (S is low) THEN F(x) is low.

IF (CR is high) AND (S is average) THEN F(x) is medium.

The categorical value for CR includes {low, average, high}, the categorical values for S include {low, average, high}. The decision attribute F(x) i.e. output has four categorical values that include {very low, low, medium, high, very high}. A total number of possible fuzzy inference rules will be 9(3*3), hence there are two linguistic states. The fuzzy rules are given in **Table 1**. An example membership plot of the input variables CR, S and output variable F(x) using triangular membership function is shown in **Figure 2**.

Defuzzification is used to get the crisp values from the fuzzy inference rules. The input fuzzy set 'μ' is defuzzified into crisp value 'c' using centroid technique.

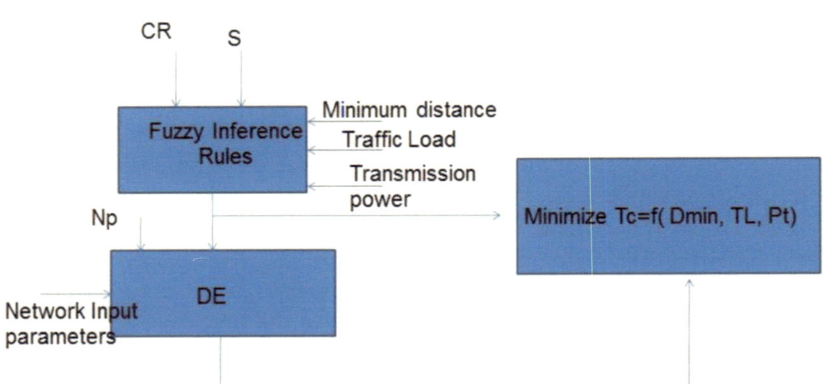

Figure 1.
Optimization control flow model using fuzzy differential evolution.

CR	**S**		
	Low	**Average**	**High**
Low	VL	L	M
Average	L	M	M
High	M	H	VH

VL, very low; L, low; medium, M; H, high.

Table 1.
Fuzzy rules-DE control inputs F(x).

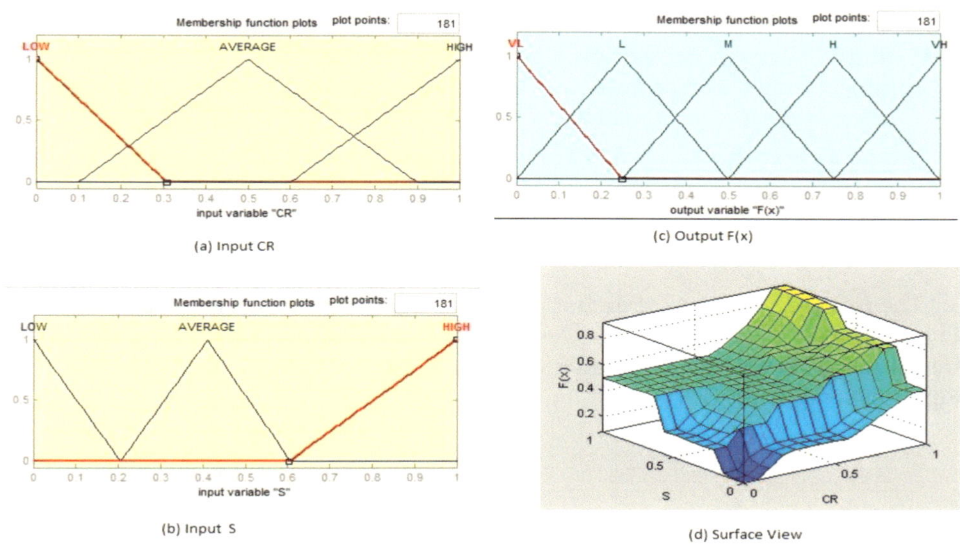

Figure 2.
Membership plot of DE inputs CR, S and output F(x) with surface view plot.

For example the linguistic values of CR (low = 0.3, average = 0.6, high = 1) and S (low = 0.2, average = 0.6, high = 1) the crisp output can be calculated by

$$C(out) = \sum xi\ \mu(xi) / \sum \mu(xi) \tag{10}$$

where xi is the CR or S and $\mu(xi)$ is the linguistic value.

As the network control parameters like the minimum distance (Dmin), Traffic Load (TL) and Transmission power (Pt) are uncertain parameters, a fuzzy inference method is used to map the input parameters with output cost metric function TC which refers only the energy consumption of mesh nodes. A sample rule set is given by:

1. IF (Dmin is short) AND (TL is low) AND (Pt is low) THEN TC is low.

2. IF (Dmin is medium) AND (TL is medium) AND (Pt is low) THEN TC is acceptable.

3. IF (Dmin is long) AND (TL is high) AND (Pt is low) THEN TC is high.

4. IF (Dmin is very long) AND (TL is high) AND (Pt is high) THEN TC is very high.

The categorical value for Dmin includes {short, medium, long, very long}, the categorical values for *TL* include {low, medium, high} and the categorical value for *Pt* includes {low, medium, high}.The decision attribute i.e. output cost has 3 categorical values that include {low, acceptable, high}. A total number of possible fuzzy inference rules will be 36(4*3*3), hence there are three linguistic states. The fuzzy rules for TC are tabulated in **Table 2.** with respect to four categorical values of minimum distance (a) short (b) medium (c) long and (d) very long. For example the linguistic values of *D*min ranges from zero to 1 km and categorized as (short = 0.2; medium = 0.4; long = 0.6, very long = 1), the values of P_t ranges from 1 to 15 dBm and categorized as (low = 5; medium = 10; high = 15). The values of *TL* is the number of clients associated to a MR which ranges from 1 to 45 and categorized for fuzzy rule as (low = 15; medium = 30; high = 45).

Input: TL➔Traffic Load; Pt➔Transmission power; Distance.

Output: TC➔Transmission Cost.

Output variables: L➔Low; A ➔Acceptable; H➔High.

The membership plots of each input variable and the output variable shown in **Figure 3(a)–(d)**. The input variables are minimum distance, traffic load and transmission power. The fuzzy rule viewer and surface view plot are shown in **Figures 4** and 5.

Pt	**TL**		
	Low	**Medium**	**High**
(a) Distance: short TC			
Low	L	L	A
Medium	L	A	H
High	A	A	H
(b) Distance: medium TC			
Low	L	L	H
Medium	L	A	H
High	A	H	H
(c) Distance: long TC			
Low	A	H	H
Medium	A	H	H
High	L	A	A
(d) Distance: very long TC			
Low	H	H	H
Medium	A	H	H
High	A	A	H

Table 2.
Fuzzy rules_Transmission cost.

DOI: http://dx.doi.org/10.5772/intechopen.83747

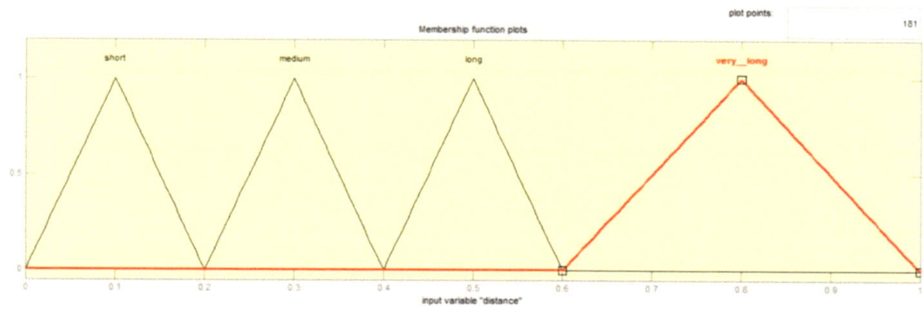

(a)

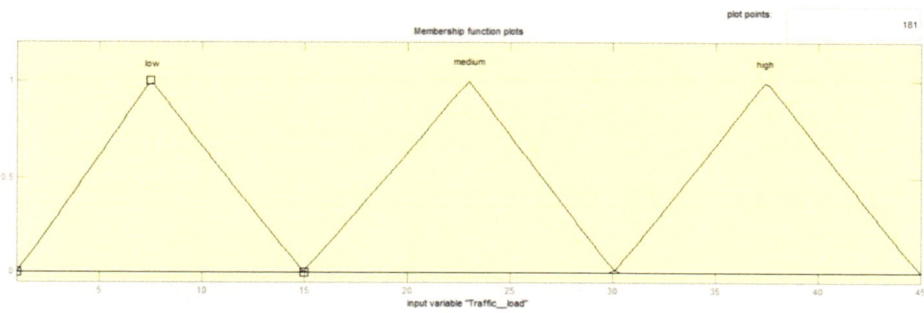

(b)

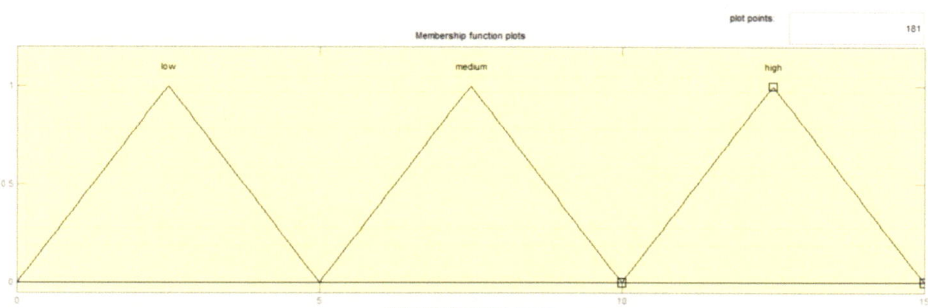

(c)

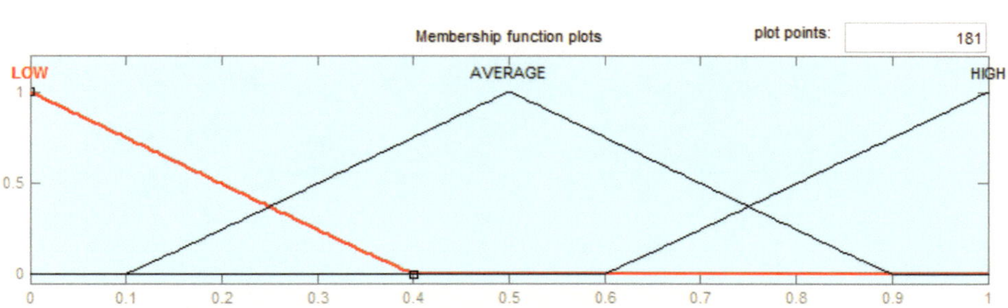

(d)

Figure 3.
Membership plot of network inputs and output transmission cost metric.

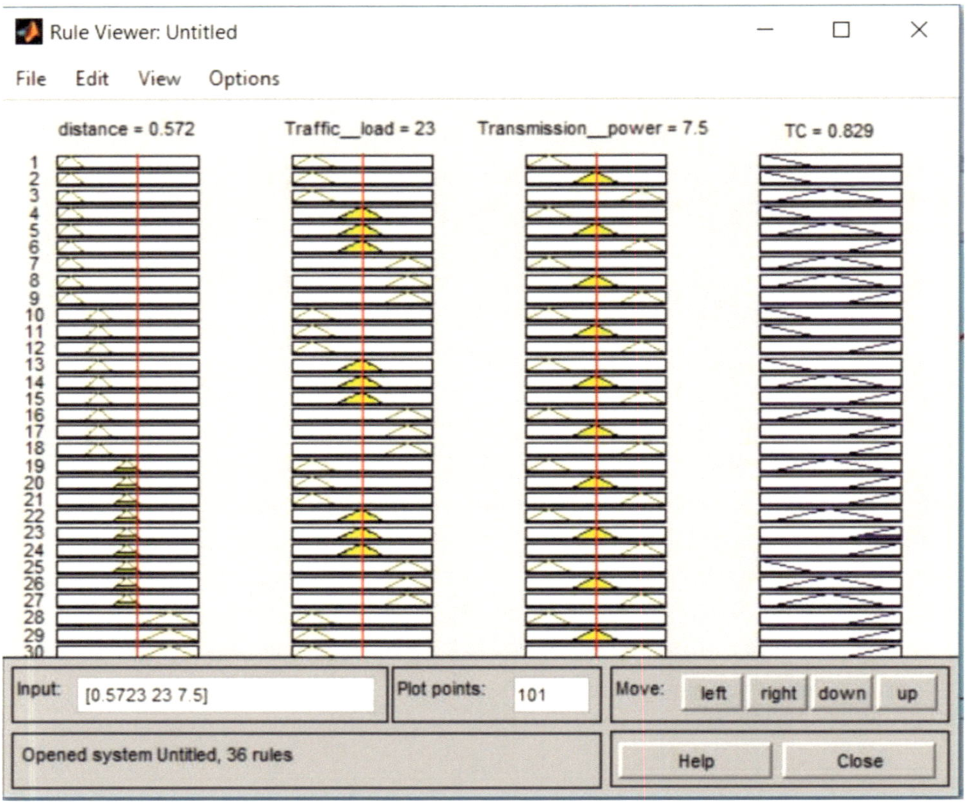

Figure 4.
Fuzzy rule viewer.

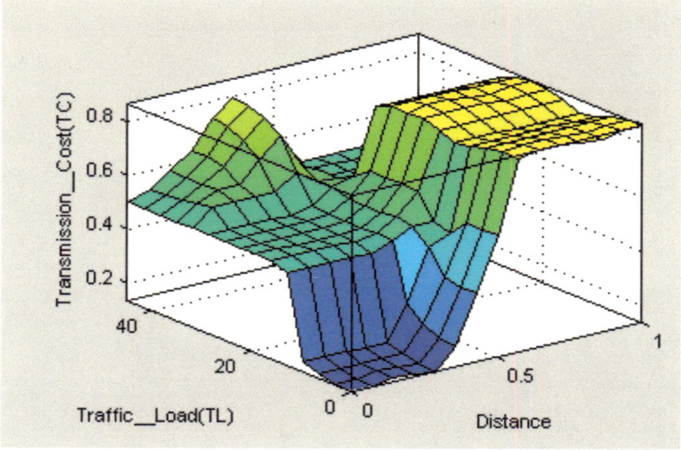

Figure 5.
Surface view plot.

5. Energy aware nearest cell association (EANCA) algorithm

An energy aware MR placement is accomplished by developing an EANCA algorithm. The conditions for associating the MC with a MR are the residual energy must be greater than the consumed energy and the power consumed by the both the mesh nodes must be less than the maximum power allocated. Even if any MR has no client association and it is in idle state, still there is some minimum amount of energy consumption. In order to overcome this, the proposed energy aware scheme turns the inactive mesh routers to sleep mode thus minimizing the energy consumption [13]. The energy consumption level of a node at any time of the simulation can be determined by finding the difference between the current energy value and initial energy value.

The methodology is illustrated as follows:

Step 1: The network input parameters like the number of mesh routers and clients are specified.

Step 2: Compute the distance of clients with respect to each router and store in a matrix and calculate the minimum distance at each generation.

Step 3: The DE control parameters CR,S are applied to the fuzzy inference engine and adaptively controlled to select the best optimum inputs for best result.

Step 4: The important parameters like distance between MR and MC, Traffic Load(TL) for each router and transmission power of router are fuzzy ruled and adaptively tuned for optimal setting.

Step 5: If TL = f(medium, high, very high) and
If Dmin = f(short, medium) and.
If Pt = f(medium, high) then MR $\in$ active mode.
If TL = f(low, very low) and.
If Dmin = f(long, very long) and.
If Pt = f(low) then MR $\in$ sleep mode.

Step 6: Check the QOS constraints

Step 7: If true then
Mc(j) $\in$ MR(i)
Er(new) = Er-1
Else if.
Mc(j) $\notin$ Mr.(i)
Then.
Compute the fitness value.

Step 9: End

The performance of the proposed scheme is analyzed based on three important metrics PDR, throughput and FR.

5.1 Performance metrics

Throughput. Network throughput is the average rate of successful message delivery over a communication channel.

Packet delivery rate. The ratio of the average number of data packets received by the destination node to the number of data packets transmitted.

Failure rate (FR). In a time slot there is a possibility for a MC not to be assigned to a MR, hence they get disconnected which is referred as connection failure. The network performance is evaluated through a metric failure rate and it is defined as the number of failures to the attempts to make the connection.

6. Simulation results and discussion

The proposed approach is evaluated in NS2 simulator and compared with the existing algorithms. The MRs is deployed in a large terrain area of 1000 m $\times$ 1000 m and the clients are distributed normally. The deployment field is equally divided into grid cells with equal area. The simulation period is set as 12 hours i.e. half a day, which is divided into 108 consecutive slots each with time duration of 400 seconds. The failure rate threshold is set as 1. The network performance metrics are evaluated for this simulation model. The simulation is repeated for 200 generations and the FR percentage is calculated after each generation. The input simulation parameters are given in **Tables 3** and **4**.

The energy consumption is evaluated for FDE approach and is compared with the DE, SA and conventional method using only the traffic weight to allot the gateway. From the simulated results shown in **Figure 6**, it is observed that in DE and FDE placement scheme the number of routers required is minimum to cover

Parameters	**Values**
Area size	1000 m × 1000 m
MAC	802.11 s
No. of mesh routers	16
No. of mesh clients	45
Application type	CBR
Packet size	1024 bytes
Transmission power	15 dBm

Table 3.
Simulation parameters.

Parameters	**SA**	**DE**	**FDE**
Placement of nodes	Random	Random	Random
Population size	100	100	100
CR	Probabilistic selection	0.5	Fuzzy rule based selection

Table 4.
Simulation settings-optimization methods.

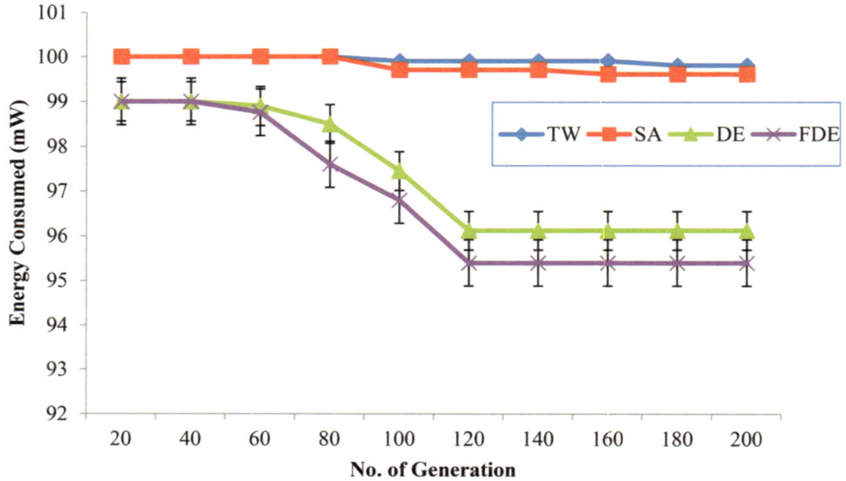

Figure 6.
Energy consumption of mesh nodes.

the given number of clients and it is found that there is a gradual decrease in energy consumption from 80th generation to 120th. From 120th generation the schemes have converged for the optimal result. Whereas SA and the conventional method using Traffic Weight (TW) allotment of gateways show high energy consumption as the number of routers required are high as well as the routers are inactive with no client association.

In order to overcome the premature convergence in DE, fuzzy DE method uses the knowledge base fuzzy rules. When CR is high than the scaling factor the convergence rate is faster. The proposed FDE scheme utilizes also the knowledge about the network load, router and client distance which enables the system to consume less power than other node placement schemes.

The FDE approach shows 5.12% lesser energy consumption than TW method, 4.4% lesser energy consumption than SA and 0.75% lesser than DE algorithm. The network performance is evaluated through metrics such as throughput and PDR. Throughput refers how much data can be transferred from one location to another in a given amount of time.

PDR is defined as the ratio between the successfully received packets in the destination to the number of data packets sent from the source node. The comparative results of the conventional methods and evolutionary approaches are tabulated in **Table 5** for 45 clients, 16 routers and one gateway in normal distribution.

The observed results show that the FDE approach produces better results compared to the other approaches. The number of failures for the clients to associate

Data rate = 12 Mbps; no. of clients = 45; no. of MRs = 16			
Methodology	**Throughput (Mbps)**	**PDR (%)**	**Failure rate (%)**
Conventional	6	61	27.35
TW	6.8	67	20.2
SA	8.9	71	13.25
DE	11	97	12.12
FDE	11.12	97.34	10.1

Table 5.
Performance evaluation.

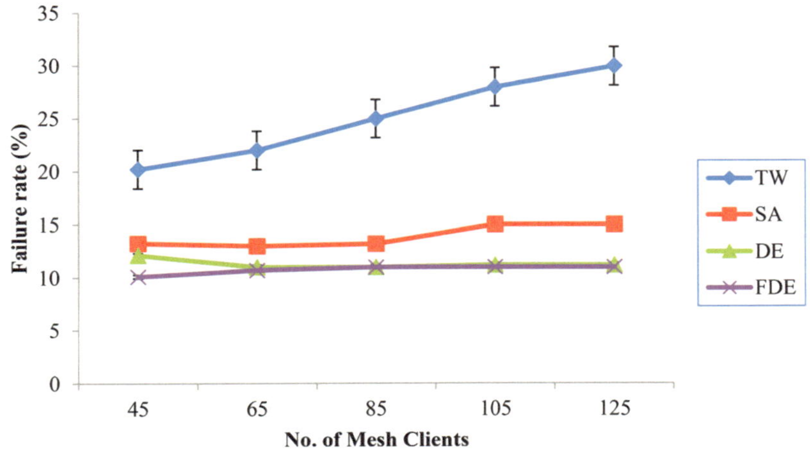

Figure 7.
Percentage of failure rate vs. no. of mesh clients.

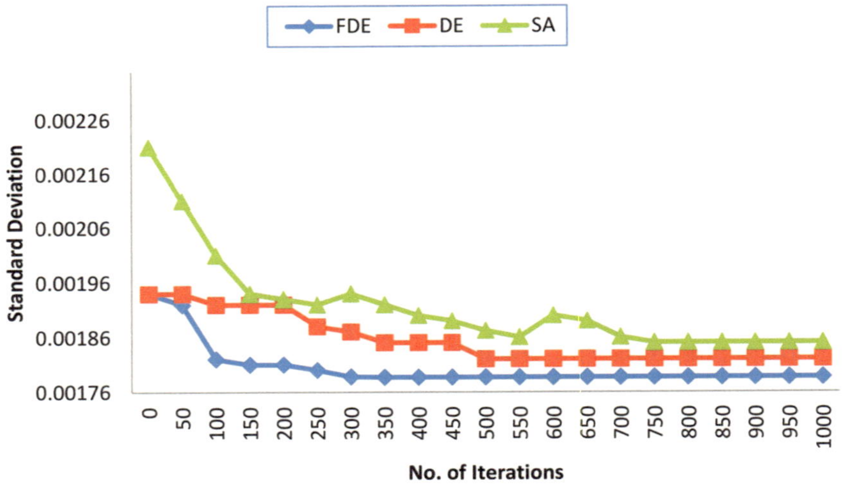

Figure 8.
Convergence rate.

with the mesh routers are high for conventional and TW methods compared to the evolutionary schemes. The results show that the FDE approach has 20% less failure rate than DE and 23.7% less than SA schemes. Even if the network size increases with more number of clients the proposed FDE approach is able to show less failure percentage as displayed in **Figure 7**.

The conventional method show a steep increase as the number of mesh clients increases whereas the evolutionary approaches tries to settle down in optimum points. As the number of clients increases FDE and DE approaches converge and show only 11% of FR. The comparison of the three evolutionary approaches with convergence graphs are shown in **Figure 8**. Standard deviation is calculated for each approach. The algorithm is run for 1000 iterations and it is observed that the result of FDE approach converged with 0.001787, which is a very low value from 300th iteration. The results obtained from FDE approach is consistent and has faster convergence speed compared to DE and SA.

7. Summary

To summarize, energy aware placement using FDE approach is proposed to minimize the energy consumption. A transmission cost metric is defined as a function. Three important parameters, the minimum distance between the MRs and MCs, the transmission power of routers and traffic load.

The deployment field is divided into cells of equal area wherein the candidate locations of each MR is positioned. Normal distribution is selected to distribute the clients as it shows 36.6% increase of PDR than SA approach in the previous module.

Usually the DE control parameters are fixed but the FDE scheme uses the CR and S values adaptively to settle for optimum point. A fuzzy inference engine is used to map the input to the output function. The uncertain network parameters are also mapped using the fuzzy inference engine to evaluate the transmission cost.

An energy aware nearest cell association algorithm is proposed to make the MRs to sleep if they are in idle state. If the MRs have no associated clients then the MR is considered to be idle. Any network device in idle state consumes power hence a sleep mechanism is introduced to place energy aware routers.

The FDE approach shows 5.12% lesser energy consumption than TW method, 4.4% lesser energy consumption than SA and 0.75% lesser than DE algorithm. The failure rate is also observed to be less than the other schemes. The proposed FDE method has 20% less failure rate than DE and 23.7% less than SA schemes. The conventional method shows a steep increase as the number of mesh clients increases whereas the evolutionary approaches tries to settle down in optimum points. As the number of clients increases FDE and DE approaches converge and show only 11% of FR. Thus the results obtained using FDE shows less energy consumption and failure rate.

Author details

G. Merlin Sheeba
Department of ETCE, Sathyabama Institute of Science and Technology, Chennai, Tamil Nadu, India

*Address all correspondence to: merlinsheebu@gmail.com

References

[1] Mamechaoui S, Senouci SM, Didi F, Pujolle G. Energy efficient management for wireless mesh networks with green routers. Mobile Networks and Applications. 2015;**20**(5):567-582

[2] Aoun B, Boutaba R, Iraqi Y, Kenward G. Gateway placement optimization in wireless mesh networks with QoS constraints. IEEE Journal on Selected Areas in Communications. 2006;**24**(11): 2127-2136

[3] Zheng Z, Zhang B, Jia X, Zhang J, Yang K. Minimum AP placement for WLAN with rate adaptation using physical interference model. In: Global Telecommunications Conference (GLOBECOM 2010); IEEE. 2010. pp. 1-5

[4] Huan X, Wang B, Mo Y. Rechargeable router placement with guaranteed QOS for green WLAN meshes networks. In: 2013 International Conference Wireless Communications and Signal Processing (WCSP). 2013. pp. 1-6

[5] Huan X, Wang B, Mo Y. Placement of rechargeable routers based on proportional fairness in green mesh networks. In: 23rd International Conference Computer Communication and Networks (ICCCN). 2014. pp. 1-8

[6] Cai LX, Poor HV, Liu Y, Luan TH, Shen X, Mark JW. Dimensioning network deployment and resource management in green mesh networks. IEEE Wireless Communications. 2011; **18**(5)

[7] Ma L, Zheng X, Lu Y, Tan X. Optimization for the deployment and transmitting power of AP based on green WLAN. In: Third International Conference on Instrumentation, Measurement, Computer, Communication and Control (IMCCC). 2013. pp. 129-134

[8] Huan X, Wang B, Mo Y, Yang LT. Rechargeable router placement based on efficiency and fairness in green wireless mesh networks. Computer Networks. 2015;**78**:83-94

[9] Storn R, Price K. Differential evolution—A simple and efficient heuristic for global optimization over continuous spaces. Journal of Global Optimization. 1997;**11**(4):341-359

[10] Merlin Sheeba G, Nachiappan A. A differential evolution based throughput optimization for gateway placement in wireless mesh networks. International Journal of Applied Engineering Research. 2014;**9**(21):5021-5027

[11] Liu J, Lampinen J. A fuzzy adaptive differential evolution algorithm. Soft Computing. 2005;**9**(6):448-462

[12] Merlin Sheeba G, Nachiappan A. Gateway placements in WMN with cost minimization and optimization using SA and DE techniques. International Journal of Pharmacy & Technology. 2015;**7**(1):8274-8281

[13] Merlin Sheeba G, Nachiappan A. Fuzzy differential evolution based gateway placements in WMN for cost optimization. Advances in Intelligent Systems and Computing SPRINGER Series. 2016;**385**:137-145

Chapter 10

Fractal and Polar Microstrip Antennas and Arrays for Wireless Communications

Paulo Fernandes da Silva Junior, Mauro Sérgio Pinto Silva Filho, Ewaldo Eder de Carvalho Santana, Paulo Henrique da Fonseca Silva, Elder Eldervitch Carneiro de Oliveira, Maciel Alves de Oliveira, Fabrício Ferreira Batista, Alexandre Jean René Serres, Raimundo Carlos Silvério Freire, Almir Souza, Silva Neto, Severino Aires de Araújo Neto and Carlos Augusto de Moraes Cruz

Abstract

This chapter presents the research done by authors in recent years on microstrip antennas and their applications in wireless sensors network. The subject is delimited to the study of conventional microstrip antennas, from which antennas with fractal and polar shapes are proposed. A detailed description of the antenna design methodology is presented for some prototypes of microstrip antennas manufactured with different dielectric substrates. Analysis of the proposed antennas has been done through computational simulation of full-wave methods. Experimental characterization of antennas and dielectric materials has been performed with the use of a vector network analyzer. The results obtained for the resonant and radiation parameters of the antennas are presented. Computer-aided design (CAD) of microstrip antennas and arrays using fractal and polar geometrical transformations results in a wide class of antenna elements with desirable and unique characteristics, such as compact, exclusive, and esthetic antenna design for multiband or broadband frequency operation with stable radiation pattern.

Keywords: fractal-shaped antennas array, polar-shaped antennas, wearable antennas, dielectric resonator antennas, wireless sensor network

1. Introduction

From the 1990s, with the advent of the Internet, the popularization of portable terminals (laptops, mobile phones, etc.) favored the telecommunications industry, and the infrastructure of networks experienced a remarkable growth [1, 2]. When the information age emerges from an increasingly networked world, the digital

IntechOpen

information and communication technology permeate the society and are increasingly important to their development [3, 4]. Modern wireless applications demand esthetic, multifunctional, portable terminals (laptops and smartphones) that operate in multiple frequency bands and can integrate different wireless services: 4G, Wi-Fi, Bluetooth, NFC, GPS, etc. Future trends toward 5G systems also require enhanced mobile broadband for emergent applications, as wireless sensors network [5].

With the rapid advance of wireless communication systems, the use of antennas in base stations and portable terminals must meet increasingly stringent criteria, such as miniaturization, integration with other systems, and multiband or broadband operation [1–4]. Due to its attractive features, low-profile microstrip antennas (MSA) and arrays are well suitable to meet the demands of fixed or mobile wireless applications [6–10].

Antenna parameter specifications change according to application. Indeed, fixed antennas must have high gain, stable radiation pattern, and bandwidth tolerance; embedded antennas should be efficient in radiation and possess larger beam width [3]. In short-range UWB wireless systems, the antenna bandwidth exceeds the lesser of 500 MHz or 20% of the center frequency [11, 12]. Thus, impedance bandwidth, gain, radiation pattern, and polarization are fundamental parameters for antenna designers to take into account.

A trend in the application of antennas for modern wireless systems is the use of compact antennas with stable radiation coverage over a wideband [2–4]. An antenna must be compact in many situations: embedded antennas, wearable antennas, camouflaged antennas, etc. However, most often an antenna electrically small narrows the impedance bandwidth, reduces gain, and limits control of the resulting radiation pattern [6, 10].

This chapter discusses the design of innovative microstrip antennas with fractal and polar shapes, which has been optimized for wireless sensors network applications. To show the advantages and disadvantages of proposed antennas, their resonant and radiation properties are compared with that presented by conventional MSAs. The antenna types addressed include patches and printed monopoles. Further developments include microstrip feeding techniques, dielectric resonator antenna (DRA), esthetic wearable antennas, and antenna arrays.

2. Microstrip antennas: types, applications, and design methodology

2.1 Types and applications

Since the concept of microstrip radiators was introduced by Deschamps in 1953, microstrip antennas only were manufactured in the 1970s with the use of the printed circuit technology (PCB) by Byron, Munson, and Howell [13–16]. Since then, microstrip antennas have been a subject of extensive research and development for military and commercial applications.

The most common type of microstrip antenna is the so-called patch antenna, which is fabricated with PCB technology by etching the shape of radiating patch above a dielectric substrate backed by a ground plane. Conventional patch shapes that result in narrowband and wide-beam antenna include square, rectangular, circular, and elliptical. Patch antennas have a low profile and can be mechanically robust and shaped to conform to the curving surfaces or embedded into portable terminals.

From the initial concept introduced in [13], a variety of MSA has been proposed to meet the operating requirements in modern wireless applications. **Figure 1** illustrates some examples of these antennas fabricated using PCB technology for different types of excitation: microstrip, CPW, coupled, and coaxial.

The operating bandwidth of an antenna is an initial design specification of paramount importance to the antenna designer. The frequency bands defined for some

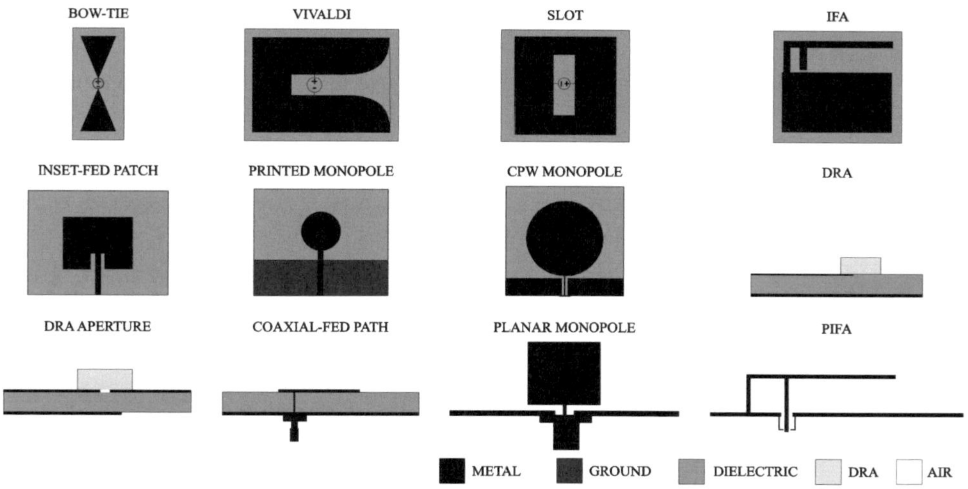

Figure 1.
Antennas manufactured using the PCB technology.

Application	*Center Frequency*	*BW (MHz)*	*BW (%)*
NFC	13.560 MHz	13.553-13.567	0.1
GPS	1575.42 MHz	1563-1587	1.5
3G (UMTS)	1900 MHz	uplink (1920-1975)	2.8
	2100 MHz	downlink (2110-2165)	2.6
4G (LTE)	700 MHz	uplink (708-748)	5.5
	(708-803 MHz)	downlink (763-803)	5.0
	2.5 GHz (2500-2690 MHz)	uplink (2500-2570)	2.7
		downlink (2570-2620)	1.9
		downlink (2620-2690)	2.6
Wi-Fi (IEEE80211b,g)	2.45 GHz	2400.0–2483.5	3.4
Wi-Fi (IEEE80211a)	5.8 GHz	5150-5350 (*indoor*)	3.8
		5470-5725	4.5
		5725-5850	2.2
UHF TV	638 MHz	470-806	52.6
UWB	6.85 GHz	3100-10600	109.5

Table 1.
Frequency bands of wireless communication services.

wireless applications are shown in **Table 1**. Conventional patch antennas suffer with narrow impedance bandwidth, low gain, and low power handling capability [6]. However, patch antennas have been applied for portable devices and base stations. A challenge for the designer is to enhance the patch antenna impedance bandwidth without compromising its radiation properties. A variety of broadband techniques for patch antennas can be found in the literature [3, 4, 9].

The microstrip antennas (IFA, Inverted-F Antenna, and PIFA, Planar Inverted-F Antenna) are widely used in wireless communication terminals [2–4]. Printed monopole antennas are very popular in ultra-wideband applications [3]. Discrete patch or monopole radiators can be arranged in versatile arrays to improve bandwidth and directivity or to synthesize a given radiation pattern [6–10].

Fractal antennas have a natural multiband behavior and compact design and can be used as a reconfigurable microstrip antenna [17–19]. Optimized fractal antennas in size and performance are suitable for wireless applications [20]. Currently, fractal antennas have several commercial applications, and international companies such as Fractal Antenna Systems, Fractus, Rayspan, and Ficosa International, among others, explore the unique properties of fractals for the manufacture of commercial antennas. Recently, polar shape commercial antennas inspired by the Gielis formula have also been proposed [21].

2.2 Microstrip antenna design methodology

In this section, a microstrip antenna design methodology using PCB technology is presented, and preliminary results are presented. After choosing a type of MSA, its design had been done in order to meet the application criteria. Often, design requirements are conflicting, for example, when a small-volume antenna with a wide bandwidth and high gain is desired.

The design of conventional patch antennas (with square, rectangular, circular shapes) is already well established [6]. In general, the initial patch dimensions are given by analytical expressions obtained from approximated models, for example, the cavity model [6].

After initial design phase, an accurate analysis of the resonant and radiation antenna properties is made through computer simulation of a full-wave method: MoM, FDTD, FEM, etc. At this step, the initial dimensions of an antenna (radiator, feed line, ground plane, etc.) can be adjusted by designer for fine tuning of resonant frequency, bandwidth, and gain, among other parameters.

A trend in the development of microstrip devices involves an intensive use of computational resources available in the application of CAD tools, computational electromagnetic analysis methods, as well as computational intelligence tools for modeling and optimization [22, 23]. **Figure 2** shows a block diagram with the main steps of the methodology used in the development of microstrip antenna prototypes.

In this design approach considered, the microstrip antennas built on single dielectric layer are fed by microstrip lines. Computational simulations of microstrip antennas were done with the use of commercial software ANSYS Designer®. A developed FDTD-3D method also is applied for antenna analysis.

MSA manufacture was performed using two PCB techniques: corrosion with iron perchloride and with a milling machine, model LPKF ProtoMat S103. **Figure 3(a)**

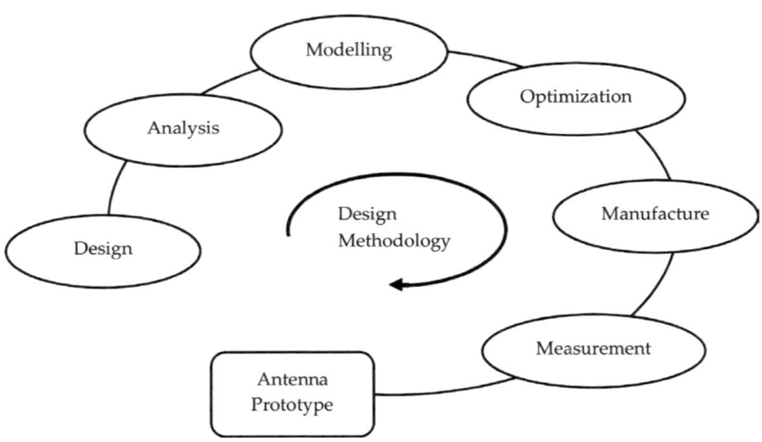

Figure 2.
Design methodology for microstrip antennas.

shows enlarged images of microstrip bends and T-junctions made using chemical corrosion and milling machine, whose manufacturing results are more accurate. Measured values of antenna parameters were obtained using a vector network analyzer (model S5071C, Agilent Technologies).

Relative dielectric permittivity and loss tangent of dielectric materials can be obtained by the following methods: coaxial probe, free space, resonant cavities, and capacitive methods [24]. The characterization of the dielectric materials (ceramic, polyamide, and denim) was performed by probe method using E5071C VNA (300 kHz–20 GHz) and Dielectric Probe 85,070 program, **Figure 3(b)**. **Figure 4(a)** shows results for ceramic dielectric. **Figure 4(b)** shows results of a compact and broadband inset-feed DRA antenna for operation in 2.4 GHz band. A list of dielectric material parameters addressed in this work is presented in **Table 2**.

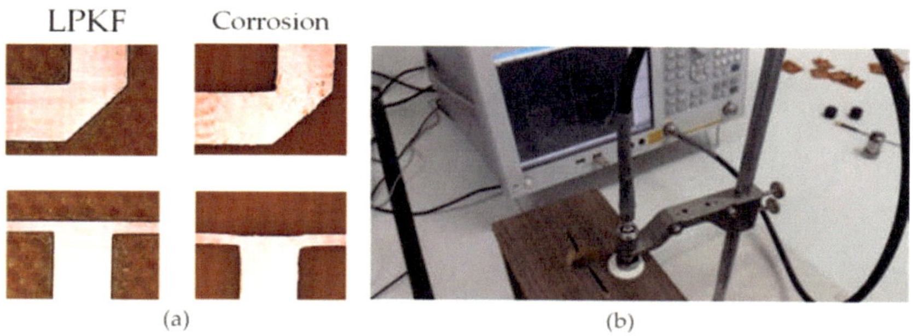

Figure 3.
(a) Images of PCB results; (b) ceramic characterization with dielectric probe 85,070.

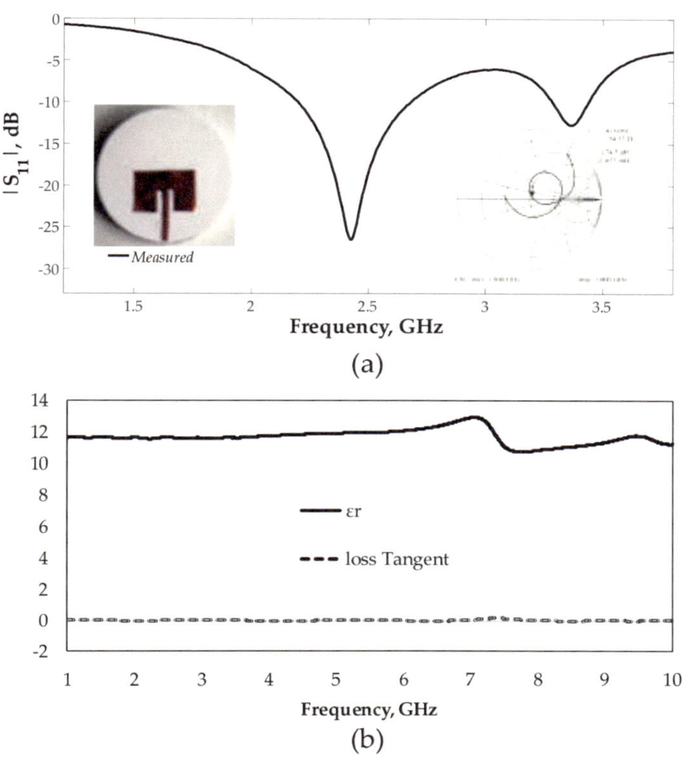

Figure 4.
DRA: (a) measured ceramic material parameter; (b) compact and broadband 2.45 GHz.

Dielectric Material	*Cost*	ε_r	*tan (δ)*	*h (mm)*
Fiberglass	low	4.40	0.002	1.55
Glass epoxy	low	5.2	0.002	1.55
Duroid	medium	2.2	0.0004	1.57
Ceramic (Calcium Titanate – $CaTiO_3$)	high	12	0.006	8
Polyamide	low	4	0.004	0.005
Denim	low	2.12	0.08	1

Table 2.
List of dielectric materials.

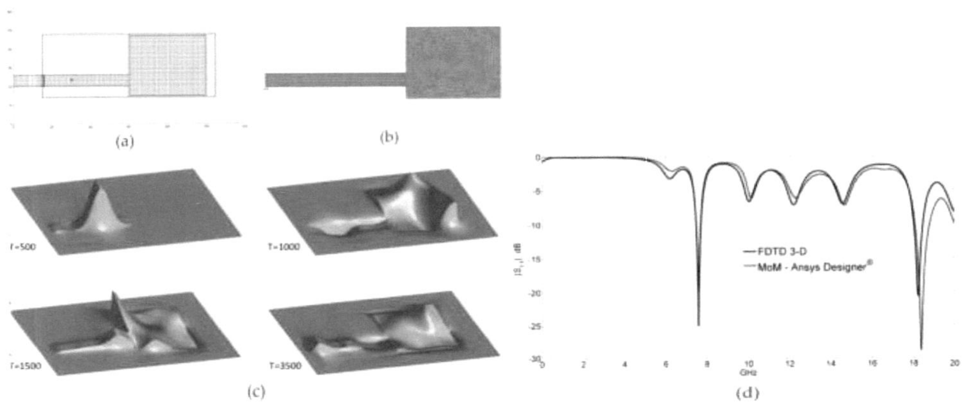

Figure 5.
Microstrip patch antenna analysis: (a) FDTD, uniform mesh; (b) MoM, tetrahedral mesh; (c) Ez-field propagation in the time domain; (d) comparison between simulation results.

Figure 5 shows analysis results for a benchmark patch antenna proposed by Sheen [25]: using a homemade FDTD-3D method, developed according to [26], and using Ansys Designer (MoM). Sheen's antenna geometry is illustrated in **Figure 5(a)** superimposed by the rectangular uniform FDTD mesh; MoM tetrahedral mesh is shown in **Figure 5(b)**.

The FDTD simulation makes it possible to observe the electromagnetic fields in the time domain. The Ez-field propagation in dielectric layer of an incident Gaussian pulse is illustrated in **Figure 5(c)**. After the occurrence of multiple reflections in the patch contours, the reflected wave back through the microstrip line is used to compute the reflection coefficient. In **Figure 5(d)** the obtained analysis results in the frequency domain are compared. The simulation time of each method depends on the computational mesh, and in this example run, it is about 15–30 minutes, which is a computing time lower than that spent in 1990 by Sheen, 12 hours [25].

Figure 6 illustrates the square patch antennas designed to operate at 2.45 GHz considering different types of microstrip line feeding techniques (direct, quarter-wave transformer, inset-fed). A combination of these feeding techniques also proposed for impedance matching of patch antenna, **Figure 6(d)**. In addition, spurline filter can be inserted into the microstrip line feed for harmonic rejection with minimum degradation of the antenna radiation pattern, **Figure 6(e)**.

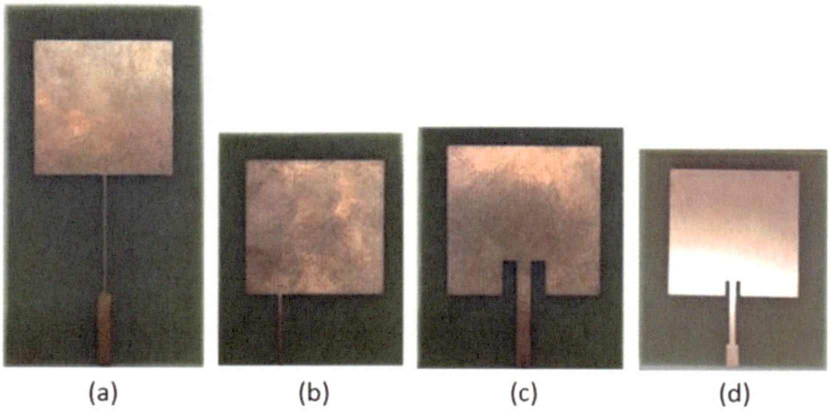

Figure 6.
Square patch antennas and microstrip line feed techniques: (a) direct, (b) QWT, (c) inset-feed, (d) hybrid, (e) hybrid with double-arm spurline filter.

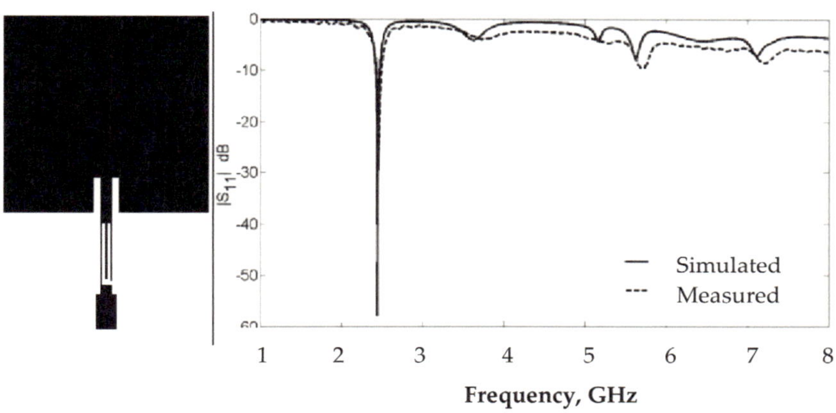

Figure 7.
(a) Layout of 2.45 GHz single-band square patch antenna, (b) simulated and measured results for reflection coefficient.

The inset-feed and QWT techniques have been combined to obtain a hybrid impedance matching with wider microstrip line section. We insert the spurline band-stop filter in QWT microstrip line section in order to suppress high-order patch resonances. **Figure 7(a)** shows layout and dimensions of such antenna considering low-cost FR-4 fiberglass dielectric substrate (see **Table 2**). **Figure 7(b)** shows simulated and measured results for proposed single-band 2.45 GHz square patch antenna.

3. Fractal and polar transformations

3.1 Types and applications

From a mathematical point of view, a fractal refers to a set in Euclidean space with specific properties, such as self-similarity or self-affinity, simple and recursive definition, fractal dimension, irregular shape, and natural appearance [27]. Fractal geometry is the study of sets with these properties, which are too irregular to be described by calculus or traditional Euclidian geometry language [27, 28].

Fractals are resorted to conventional classes, such as geometrical fractals, algebraic fractals, and stochastic fractals [29]. Two common methods used to generate mathematical fractals are iterated function systems (IFS) and Lindenmayer systems [27–30].

IFS method used to generate a 2-D fractal consisting of a collection of affine transformations with probability given by (1). Affine transformations are most commonly used in IFS. The coefficients of a two-dimensional affine transformation represent the IFS code for scaling, rotations, and translations. An affine transformation of a point to the point is given in (2).

$$\begin{cases} T_1: & (a_1,\ b_1,\ c_1,\ d_1,\ e_1,\ f_1,\ P_1) \\ T_2: & (a_2,\ b_2,\ c_2,\ d_2,\ e_2,\ f_2,\ P_2) \\ \vdots & \vdots \\ T_m: & (a_m,\ b_m,\ c_m,\ d_m,\ e_m,\ f_m,\ P_m) \end{cases} \quad (1)$$

$$\begin{cases} x_{n+1} = ax_n + by_n + e \\ y_{n+1} = cx_n + dy_n + f \end{cases} \quad (2)$$

IFS algorithm consists of four steps: (i) start with an arbitrary point in the plane p0 = (x0, y0); (ii) pick a random transformation, Tm, according to the probabilities, Pm; (iii) transform the point p1 = Tm (p0) and plot it; and (iv) go to step 2. IFS algorithm is continued ad infinitum (for ideal fractal) or until a given number of fractal iterations is reached (for pre-fractals).

Lindenmayer system (or L-system) was initially conceived to model growth phenomena in biological organisms [31]. An L-system grammar handles an initial string of symbols (axiom) and includes a set of production rules that may be applied to the symbols (letters of the L-system alphabet) to generate new strings. A graphic interpretation of strings, based on turtle geometry, is described in [29, 32]. A state of the turtle is defined as a triplet (xk, yk, φk) where coordinates (xk, yk, φk) and angle φk represent the turtle's position and direction, respectively, (3).

$$\begin{cases} x_{k+1} = x_k + d\cos(\varphi_k) \\ y_{k+1} = x_k + d\sin(\varphi_k) \end{cases} \quad (3)$$

The simplest class of L-systems is termed deterministic and context-free or DOL-systems [29, 32]. DOL-system is defined as a triple $H = (V, \omega, \Pi)$, where V is the L-system alphabet, ω is the axiom word, and Π is a finite set of productions. Formal definitions of D0L-systems and their operation can be found in [29, 32]. Given the initial state of turtle (x_0, y_0, φ_0), step size d, and the angle increment $\Delta\varphi$, the turtle can respond to the commands in L-system strings $v \in V$ and represented by the following symbols [29, 32]:

F → Move forward a step of length d, and change state of the turtle according to (3). A line segment between points (x_k, y_k) and (x_{k+1}, y_{k+1}) is drawn.

f → Move forward a step d without drawing a line. The turtle state changes as above.

+ → Turn right by angle $\Delta\varphi$. The next state of the turtle is given by $(x_k, y_k, \Delta_k + \varphi_k)$.

- → Turn left by angle $\Delta\varphi$. The next state of the turtle is given by $(x_k, y_k, \varphi_k - \Delta_k)$.

In **Figure 8**, four examples of fractal iterations using IFS and L-system are shown.

Like fractals, polar transformations give rise to a wide class of shapes. A polar transformation is defined in this chapter through a vector function $\vec{v}(t) = (x(t), y(t)),\ t \geq 0$,

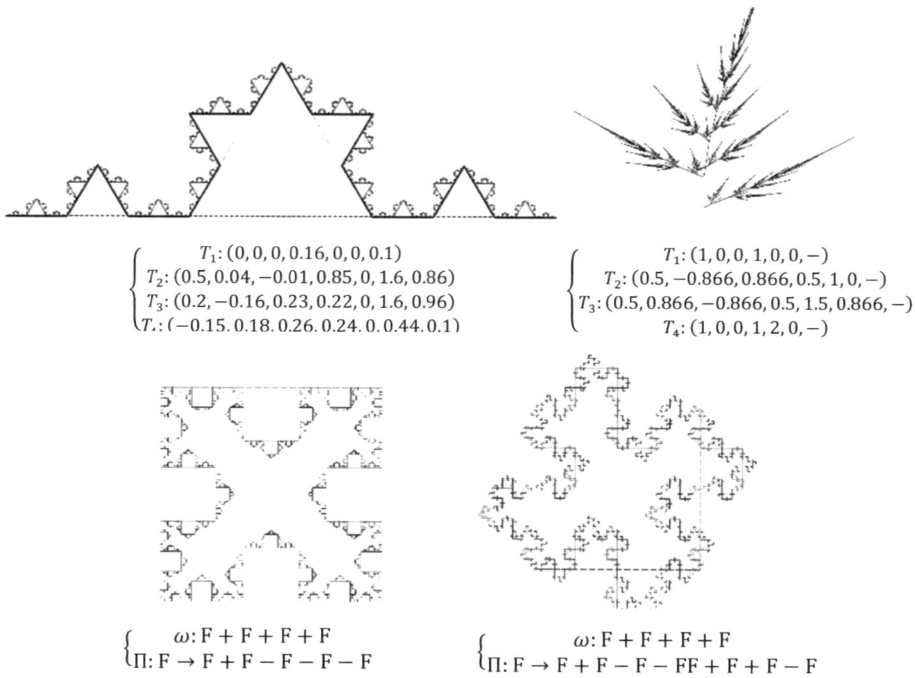

Figure 8.
IFS and L-system pre-fractals: (a) Koch curve; (b) modified Barnsley fern; (c) Koch Island; (d) Minkowski Island.

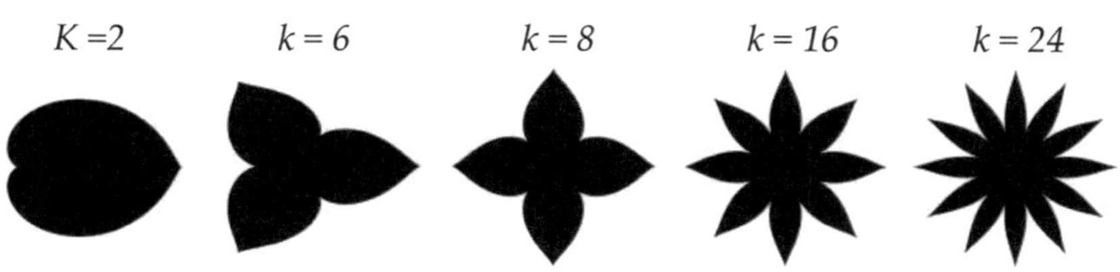

Figure 9.
Esthetic polar transformation for k varying up to k = 32 petals in (5).

that is, for each real value, t is associated with a vector in $\Re^2$, (4). An example of an esthetic polar transformation defined by (5) is presented in **Figure 9** for k varying up to $k = 32$ petals.

$$\begin{aligned} \vec{v}(t) &: I \to \mathfrak{R}^n \\ t &\to \vec{v}(t) \end{aligned} \tag{4}$$

$$\vec{v}(t) = \left(1 + \frac{\cos(t)}{2}\right) \cdot \left(\cos\left(\frac{2t - sen(2t)}{k}\right), \sin\left(\frac{2t - sen(2t)}{k}\right)\right), \quad 0 \le t \le k\pi \tag{5}$$

4. Fractal and polar-shaped microstrip antenna

4.1 Koch fractal microstrip antenna

The design of pre-fractals patch antennas has been a subject of great interest to designers and researchers in the field of antennas. Previously published

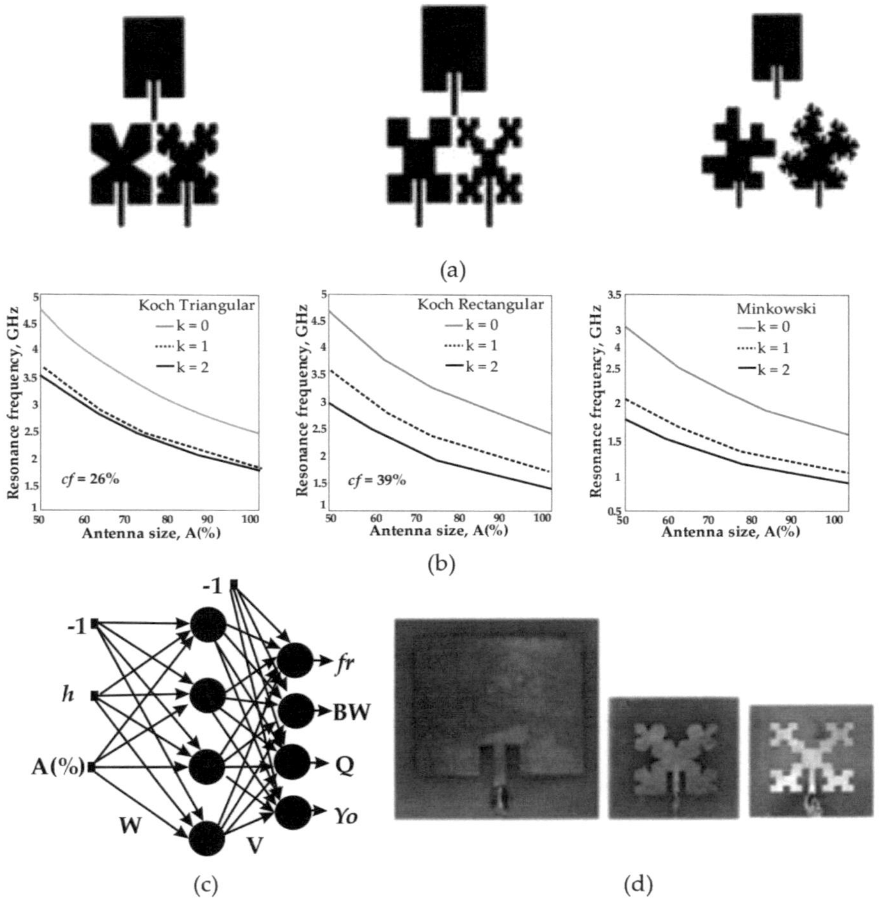

Figure 10.
Pre-fractals patch antennas: (a) image layouts; (b) parametric analysis; (c) neuromodeling; (d) image comparing overall sizes of the built patch antennas—Rectangular and pre-fractals.

works by the authors have contributed to this research area, showing the miniaturization of inset-fed patch antennas with the use of Koch and Minkowski pre-fractals [13, 19, 33], **Figure 10(a)**. Frequency compression factors of 26.1, 39, and 42% were observed for level 2 pre-fractals: triangular Koch, rectangular Koch, and Minkowski, respectively [13, 19, 33]. Pre-fractal patch antennas are defined with two fractal parameters: iteration number (level) and scaling factor. They possess a large design region of interest, **Figure 10(b)**; are easy to model using neural network, **Figure 10(c)**, [19]; and their shapes and multiband behavior facilitate frequency reconfiguration [5]. The unique properties of geometric fractals are useful to synthesis of more compact patch antennas, **Figure 10(d)**, [13, 19, 33, 34].

4.2 Wearable teragon antennas

The use of wearable antennas is necessary that have some characteristics as: easy interaction with the body, low visual impact, preferably low cost, and flexible structure [19]; for this reason, the materials used in the manufacture of the wearable antennas must follow some requirements: easy interaction with the body, flexible structure, reduced visual impact, and preferably low cost [19].

Teragon was a term coined by Mandelbrot that literally means, "monster curve" [28]. The proposed wearable teragon patch antennas are based on a square patch antenna with displaced microstrip line feed. Square patch antenna dimensions are calculated according to [13, 28, 34]. Pre-fractal teragons were developed with a scale factor of R = 6 and number of copies, n = 18. **Figure 11(a)** shows dimensions and shapes of the teragons. Images of built antenna prototypes with polyamide flexible dielectric substrate are shown in **Figure 11(b)**. Obtained simulated and measured results for reflection coefficient and gain are shown in **Figure 12**.

Figure 12a shows the comparison between simulated and measured reflection coefficient of the wearable flexible antennas. The increase of the patch perimeter by the use of teragon shapes provides a reduction of the resonant frequencies. The main highlight is for teragon 1, with reduction of approximately 1 GHz, when compared to the initial square patch antenna.

The gain (dB) simulated in resonant frequencies of the wearable path antennas is shown in **Figure 12b**. As noted, the gain is reduced when fractal level increase. The initial square patch antenna presented higher gain, with maximum gain in end-fire direction of 6.13 dBi, and the teragon 1 showed the maximum gain of 4.26 dBi (**Figure 12b**).

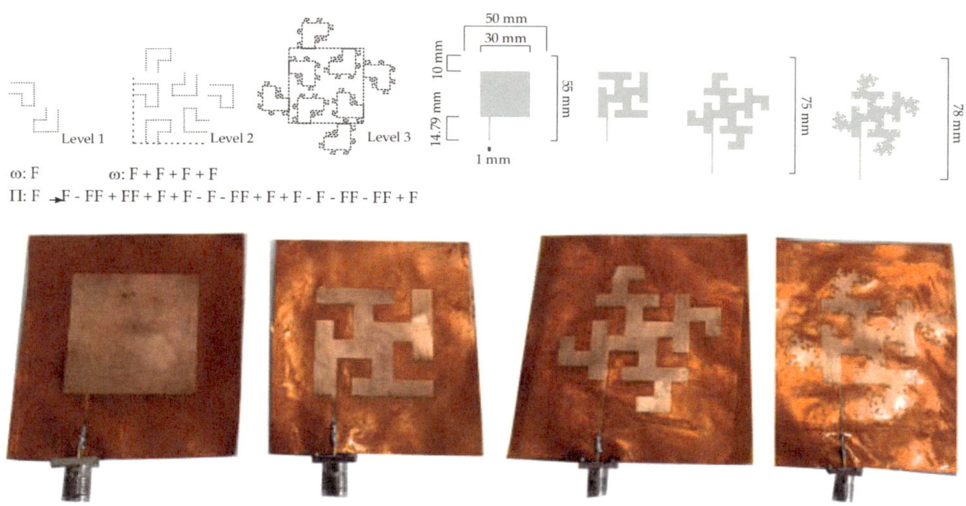

Figure 11.
Pre-fractal teragon patch antenna design steps (a) Matlab dxf images, (b) layouts, (c) prototypes.

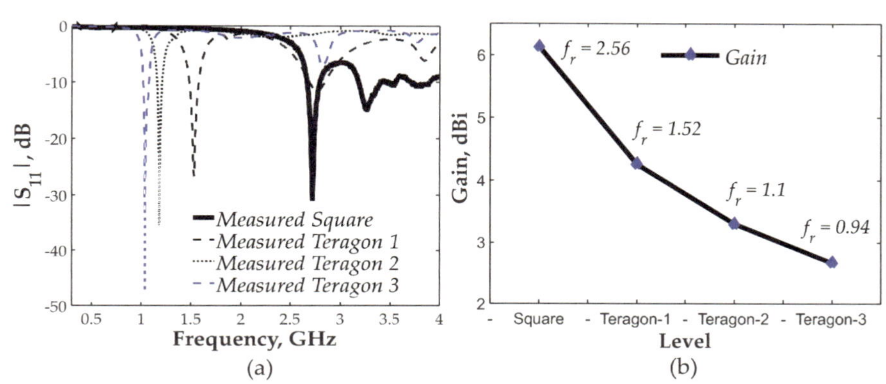

Figure 12.
Results of teragon antenna: (a) measured, (b) gain in dBi.

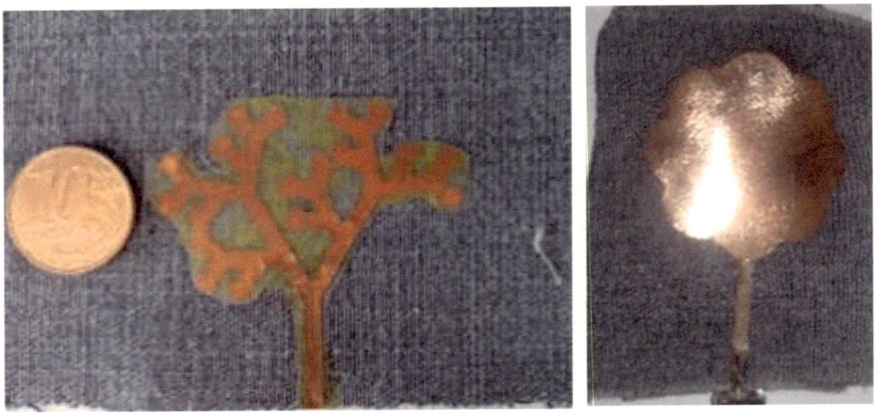

(a) (b)

Figure 13.
Wearable textile antennas: (a) L-systems, (b) polar transformer.

Several shapes were used in development of the microstrip antennas; the polar transformer is the possibility in this case. **Figure 13** shows the wearable textile antennas: patch generated by L-systems (**Figure 13(a)**) and printed monopole generated by polar transformer, **Figure 13(b)**. Printed monopole antennas (PMA) with polar shape can be observed in several works, operating mainly in the ultra-wideband (UWB), but with projects for 2G, 3G, and 4G technology and X band [35–39]. The altering frequency provided by polar shapes was observed in [38, 40], similar to the observed pre-fractal geometry applied to the PMA [41].

4.3 Polar microstrip antenna

Figure 14 shows frequency resonance of polar microstrip antenna for k-interactions (k = 1, 8 12, 16, 24, 32, 40, 48, 56, 64) and the comparison of measured

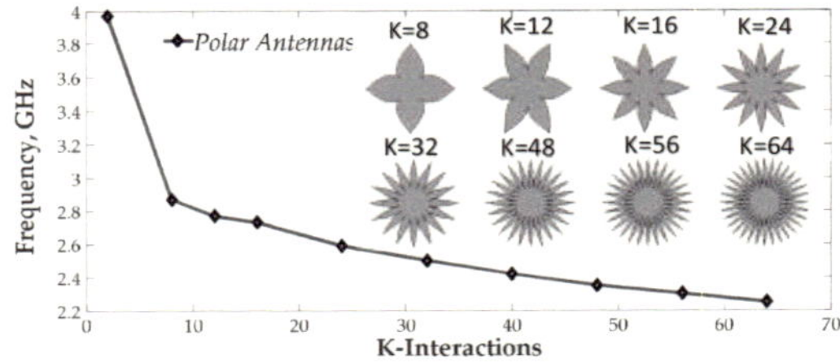

(a)

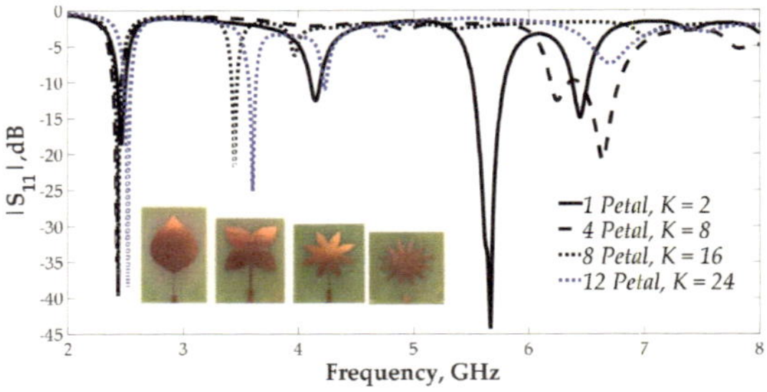

(b)

Figure 14.
Interactions of polar microstrip patch antenna: a) comparison of frequency resonance simulated; b) comparison of measured antennas.

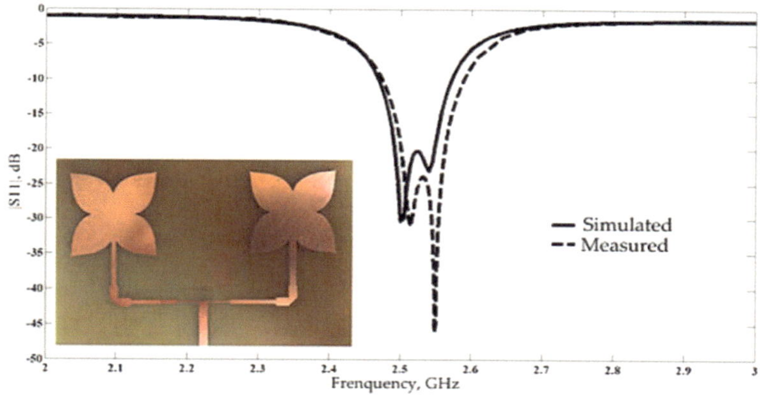

Figure 15.
Polar microstrip patch antenna array with two elements.

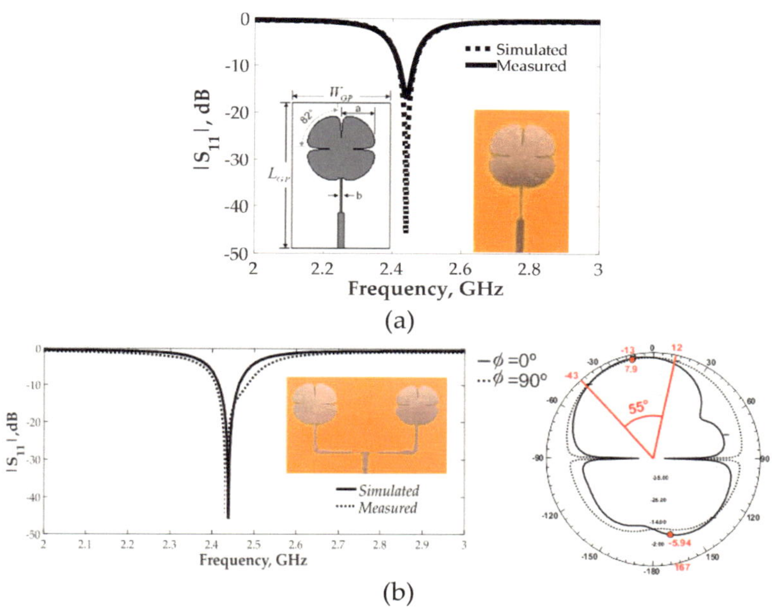

Figure 16.
Polar leaf clover patch antennas: (a) four petals, (b) array of two elements with four petals.

|S11| parameter, with prototype images. The proposed polar patch antennas are based on a circle patch antenna with displaced microstrip line feed, and quarter-wave transformer, with dimensions calculated according [7, 9].

Figure 14 shows the |S11| parameters measured by the polar antennas to k = 2, 8, 16, 24. We noted than the increase of the patch perimeter by the use of polar interaction provides a reduction of the resonant frequencies, similar to the fractal comportment. The greater difference can be observed in k = 2 and k = 8, of 3.4 GHz, and all structures with dual-frequency resonances.

Figure 15 shows the use of polar transformer in the development of the array patch antenna with 4 petals, k = 8 interactions. The polar array presented good response, with simulated and measured results closed, had loss return less than −45 dB and bandwidth of 101 MHz, and covered the WLAN band in 2.4 GHz.

The other shape used was the leaf clover, generated by (8). **Figure 16** shows the comparison of $|S_{11}|$ parameter measured and simulated with leaf clover with four and six petals for one and two patch elements and prototype images; **Table 3** presents the dimensions used. The polar antenna with six petals presented great bandwidth (52 MHz) than the polar antenna with four petals (42 MHz) and best loss return (−26.3 dB), **Figure 13(b)**.

$$r = 4.4 - \min\left(abs\left(\tan\left(2t + \pi/1\right)\right)/10, 3\right) \tag{6}$$

From the leaf clover antennas, polar array patch antennas with two and four elements have been developed. **Figure 16** shows polar array patch antennas with the shape of clover of four and six petals, with two and four elements, operating in WLAN range. The antennas presented measured bandwidth of 81 MHz and half power beamwidth (HPBW) of 55°, the inclination of radiation pattern indicating the great element used in the patch array (**Figure 17**).

Figure 18 shows the Koch fractal patch antenna array with two elements of the square geometry until Koch level 2. The applications of Koch fractal in the array

Antenna	L_{GP}	W_{GP}	*a*	*b*
4 Petals	58,83	38,22	14,12	0,56
6 Petals	58,83	38,22	14,15	0,44

Table 3.
Dimensions of leaf clover antenna (mm).

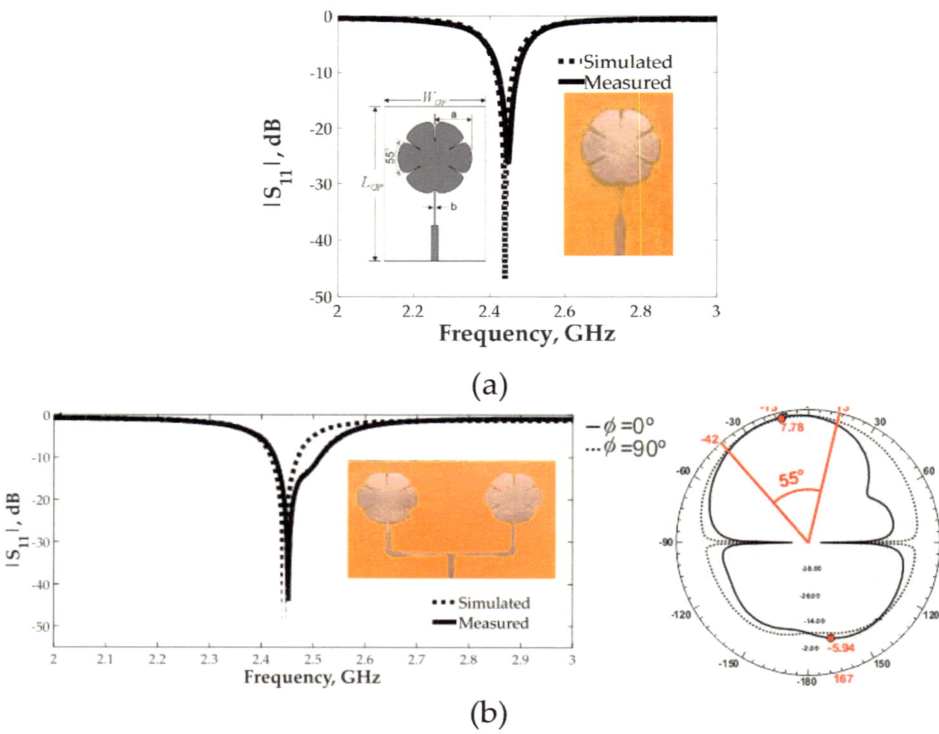

Figure 17.
Polar leaf clover patch antennas: (a) six petals, (b) array of two elements with six petals.

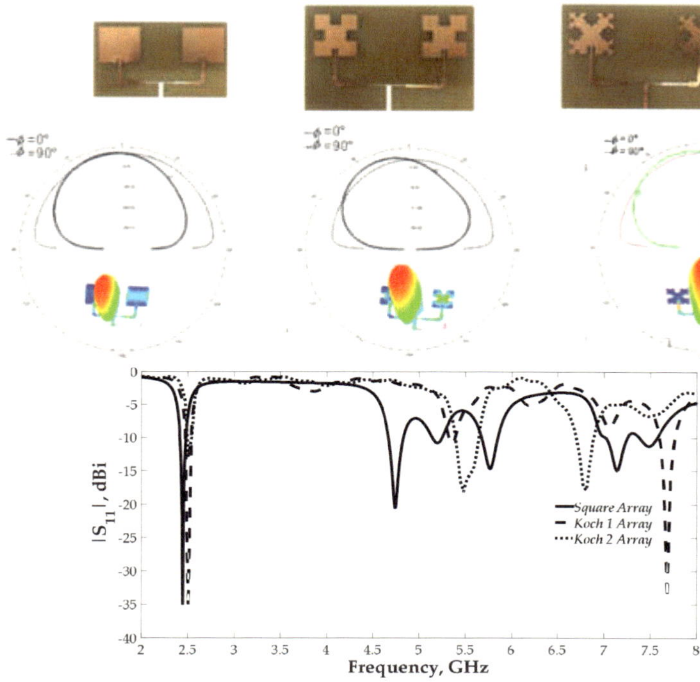

Figure 18.
Koch array patch antenna with two elements.

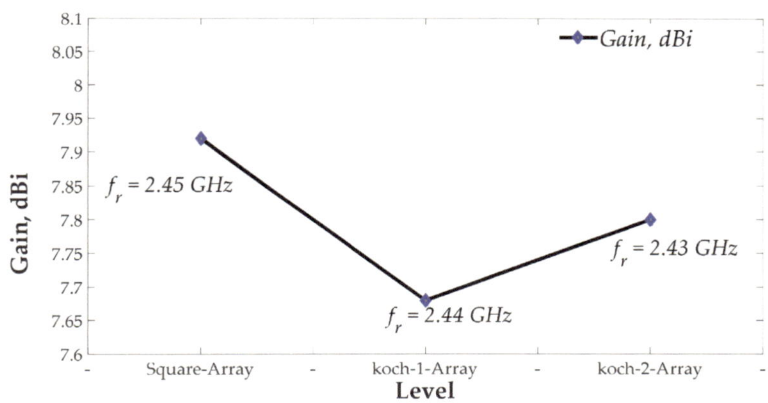

Figure 19.
Gain comparison of the Koch array patch antenna with two elements.

structure provide great bandwidth (113 MHz) and maximum gain in end-fire direction of 7.93 dBi (**Figure 19**), with variation of radiation pattern, indicating lager patch element.

5. Conclusions

In this chapter, we have described some trends for the computer-aided design of microstrip antennas (patches and printed monopoles) for wireless sensors network applications. With the use of such CAD tools, innovative designs of antennas and arrays with pre-fractals and polar motifs were approached and their properties checked. The methods of analysis, manufacturing, and measurement have been presented considering different dielectric materials (rigid and flexible) for the manufacture of the antennas. The proposed antennas have been fed by microstrip line, and different feeding techniques have been considered for matching impedances and suppression of harmonic frequencies. The unique properties of space-filling and self-similarity naturally result in more compact and multiband behavior antennas. On the other hand, it is verified from the presented results that the polar elements (like a Rosacea of n-petals) also present the property of space-filling, resulting in more compact antennas. On the other hand, pre-fractals and polar patch antennas generally have their gain and/or bandwidth reduced as the number of iterations increases, which in many wireless applications are undesirable characteristics. To overcome these limitations, we proposed the design of fractals and polar arrangements with dissimilar elements, which allows increasing the bandwidth and gain of simple antennas. Further developments included the design of printed monopole antennas for ultra-wideband applications. Flexible substrates (polyamide and denim) were used in the design of wearable antennas with esthetic appeal. The microstrip antennas with pre-fractals and polar elements have few design variables and smooth responses in the region of interest, which facilitates all steps of the design methodology.

Acknowledgements

This work was supported by CNPq under contracts 472098/2013-6 and 308509/2015-3, by the Federal Institute of Paraíba (IFPB), Federal University of Campina Grande (UFCG), State University of Paraíba (UEPB), State University of Maranhão, and FAPEMA.

Author details

Paulo Fernandes da Silva Junior[1*], Mauro Sérgio Pinto Silva Filho[1], Ewaldo Eder de Carvalho Santana[1], Paulo Henrique da Fonseca Silva[2], Elder Eldervitch Carneiro de Oliveira[3], Maciel Alves de Oliveira[4], Fabrício Ferreira Batista[4], Alexandre Jean René Serres[4], Raimundo Carlos Silvério Freire[4], Almir Souza, Silva Neto[5], Severino Aires de Araújo Neto[6]and Carlos Augusto de Moraes Cruz[7]

1 State University of Maranhão, São Luís, Brazil

2 Federal Institute of Paraíba, João Pessoa, Brazil

3 Estadual University of Paraíba, João Pessoa, Brazil

4 Federal University of Campina Grande, Campina Grande, Brazil

5 Federal Institute of Education, Science and Technology of Maranhão, São Luís, Brazil

6 Federal University of Paraíba, João Pessoa, Brazil

7 Federal University of Amazonas, Manaus, Brazil

*Address all correspondence to: paulo.junior@ee.ufcg.edu.br

References

[1] Elfergani I, Hussaini AS, Rodriguez J, Abd-Alhameed R. Antenna Fundamentals for Legacy Mobile Applications and beyond. USA: Springer; 2018. DOI: 10.1007/978-3-319-63967-3

[2] LeMoyne R, Mastroinanni T. Wearable and Wireless Systems for Healthcare I. Singapore: Springer; 2018. DOI: 10.1007/978-981-10-5684-0

[3] Chen ZN, Chia MYW. Broadband Planar Antennas: Design and Applications. Chichaster, England: John Wiley & Sons, Ltd; 2006

[4] Peres A. Wi-Fi Integration to the 4G Mobile Network. Hoboken, USA: John Wiley & Sons; 2018

[5] Yang Y, Shi JXG, Wang CX. 5G Wireless Systems: Simulations and Evaluation Techniques. Switzerland: Springer; 2018. DOI: 10.1007/978-3-319-61869-2

[6] http://www.4gamericas.org/files/6514/3930/9262/4G_Americas_5G_Spectrum_Recommendations_White_Paper.pdf

[7] Balanis CA. Antenna Theory. 3rd ed. Arizona: Wiley; 2009. p. 941

[8] Garg R, Bhartia P, Bahl I, Ittipiboon A. Microstrip Antenna Design Handbook. Boston, USA: Artech House; 2001

[9] Stutzman WL. Antenna Theory and Design. 2nd ed. New York: John Wiley & Sons; 1998. p. 598

[10] Kumar G, Ray KP. Broadband Microstrip Antennas. Boston, Mass, USA: Artech House; 2003

[11] Milligan TA. Modern Antenna Design. 2nd ed. New Jersey, USA: John Wiley & Sons, Inc.; 2005

[12] Federal Communications Commission. Revision of Part 15 of the Commission's Rules Regarding Ultra-Wideband Transmission Systems, First Report and Order (ET Docket 98-153), Adopted Feb. 14, 2002, Released Apr. 22, 2002

[13] Oliveira EEC, Silva PHF, Campos ALPS, d'Assunção AG. Small-size quasi-fractal patch antenna using the Minkowski curve. Microwave and Optical Technology Letters. 2010;**52**(4):805-809. DOI: 10.1002/mop

[14] Deschamps G, Sichak W. Microstrip microwave antennas. In: Proceedings of the Third Symposium on the USAF Antenna Research and Development Program; 1953. pp. 18-22

[15] Byron EV. A new flush-mounted antenna element for phased array application. In: Proc. Phased-Array Antenna Symp.; 1970. pp. 87-192

[16] Munson RE. Conformal microstrip antennas and microstrip phased arrays. IEEE Transactions on Antennas and Propagation. 1974;**22**(1):74-78. DOI: 10.1109/TAP.1974.1140723

[17] Chong C, Watanabe F, Inamura H. Potential of UWB technology for the next generation wireless communications. In: IEEE Ninth International Symposium on Spread Spectrum Techniques and Applications; 2006. pp. 422-429. 10.1109/ISSSTA.2006.311807

[18] Cohen N. Fractal antenna applications in wireless telecommunications. Proceedings of Electronics Industries Forum of New England. 1997:43-49

[19] Oliveira EEC, Silva PHF, Campos ALPS, Silva SG. Overall size antenna reduction using fractal elements. Microwave and Optical Technology

Letters. 2009;**51**(3):671-675. DOI: 10.1002/mop

[20] Howell JQ. Microstrip antennas. IEEE Transactions on Antennas and Propagation. 1974;**23**(1):90-93. DOI: 10.1109/TAP.1975.1141009

[21] Werner DH, Ganguly S. An overview of fractal antenna engineering research. IEEE Antennas and Propagation Magazine. 2003;**45**(1):38-57. DOI: 10.1109/MAP.2003.1189650

[22] The Antenna Company International. [Internet]. 2010. Available from: http://www.antennacompany.com [Accessed: 11 August 2015]

[23] Silva MR, Nóbrega CL, SILVA PHF, D'Assuncao AG. Optimal design of frequency selective surfaces with fractal motifs. IET Microwaves, Antennas and Propagation. 2014;**1**(9):1-5. DOI: 10.1049/iet-map.2013.0462

[24] Silva MR, Nóbrega CL, SILVA PHF, D'Assuncao AG. Optimization of FSS with Sierpinski island fractal elements using population-based search algorithms and MLP neural network. Microwave and Optical Technology Letters. 2014;**56**(4):827-831. DOI: 10.1002/mop.28214

[25] Kassem H, Vigneras V, Lunet G. Characterization techniques for materials' properties measurement. In: Minin I, editor. Microwave and Millimeter Wave Technologies from PhotonicBandgap Devices to Antenna and Applications. Rijeka: InTech; 2010. pp. 289-314

[26] Sheen DM, Ali SM, Abouzahra MD, Kong JA. Application of the three-dimensional finite-difference time-domain method to the analysis of planar microstrip circuits. IEEE Transactions on Microwave Theory and Techniques. 1990;**38**(7):849-857. DOI: 10.1109/22.55775

[27] Sullivan DM. Electromagnetic Simulation Using the FDTD Method. 2nd ed. Piscataway: IEEE Press; 2013. 182 p

[28] Mandelbrot BB. The Fractal Geometry of Nature. 3rd ed. Nova York: W. H. Freeman and Co.; 1982. p. 468

[29] Falconer K. Fractal Geometry: Mathematical Foundations and Application. 2nd ed. Londres: Wiley; 2003. p. 337

[30] Mishra J, Mishra S. L-Systems Fractals. Amsterdam, Netherlands: Elsevier; 2007. p. 274

[31] Barnsley M. Fractals Everywhere. San Diego: Academic Press; 1988. p. 394

[32] Lindenmayer A. Mathematical models for cellular interaction in development, parts I and II. Journal of Theoretical Biology. 1968;**18**(3):280-315. DOI: 10.1016/0022-5193(68)90080-5

[33] Oliveira EEC, Campos ALPS, Silva PHF. Quasi-Fractal Koch Triangular Antenna. In: 2009 SBMO/IEEE MTT-S International Microwave and Optoelectronics Conference (IMOC); 3-6 November; Belém. IEEE; 2009. pp. 163-166. DOI: 10.1109/IMOC.2009.5427607

[34] Silva PHF, Oliveira EEC, d'Assunção AG. Using a multilayer perceptrons for accurate modeling of quasi-fractal patch antennas. In: 2010 International Workshop on Antenna Technology (iWAT) Lisbon; 2010. pp. 1-4. DOI: 10.1109/IWAT.2010.5464782

[35] da Silva PF, Freire RCS, Serres AJR, Silva PHDF, e Silva JC. Bio-inspired antenna for UWB systems. In: 2016 1st International Symposium on Instrumentation Systems, Circuits and Transducers (INSCIT); Belo Horizonte; 2016; pp. 153-157. DOI: 10.1109/INSCIT.2016.7598210

[36] de Moura LCM, Cruz JDN, da Costa AP, Silva PHDF, e Silva JC. UWB cotton leaf design microstrip-fed printed monopole antenna. In: 2015 SBMO/IEEE MTT-S International Microwave and Optoelectronics Conference (IMOC) Porto de Galinhas; 2015. pp. 1-4. DOI: 10.1109/IMOC.2015.7369155

[37] Lemos NA, Silva AN, Paiva HF, Silva PHF. Four-Leaf Clover UWB Planar Monopole Antenna, MOMAG 2014: 16 SBMO. Brazil, 1 CD. 2014

[38] Silva Junior PF, Freire RCS, Serres AJR, Silva PHF, Silva JC. Wearable textile bioinspired antenna for 2G, 3G and 4G systems. Microwave and Optical Technology Letters. 2016;**58**(12): 2818-2823. DOI: 10.1002/mop

[39] Silva PF Jr, Silva PHDF, Serres AJR, Silva JC, Freire RCS. Bio-inspired design of directional leaf-shaped printed monopole antennas for 4G 700 MHz band. Microwave and Optical Technology Letters. 2016;**58**(12): 1529-1533. DOI: 10.1002/mop.29853

[40] Silva Júnior PF, Serres AJR, Freire RCS, Serres GKF, Gurjão EC, Carvalho JN, et al. Bio-inspired wearable antennas. In: Ortiz JH, editor. Wearable Technologies. Rijeka: IntechOpen; 2018. DOI: 10.5772/intechopen.75912

[41] Silva MR, Nóbrega CL, Silva PHF, D'Assunção AG. A new configuration of planar monopole quasi-fractal antenna for wireless communications. In: Digests of the 2010 14th Biennial IEEE Conference on Electromagnetic Field Computation; 9-12 May; Chicago. IEEE; 2010. pp. 1-1. DOI: 10.1109/CEFC.2010.5481786